中华思想文化术语传播工程

Key Concepts in Chinese Thought and Culture

北京历史文化名城保护关键词

汉英对照

北京历史文化名城保护委员会办公室 编

Conservation of the Historic City of Beijing: Key Terms

(Chinese-English)

外语教学与研究出版社
FOREIGN LANGUAGE TEACHING AND RESEARCH PRESS
北京 BEIJING

图书在版编目（CIP）数据

北京历史文化名城保护关键词 ：汉英对照 / 北京历史文化名城保护委员会办公室编. -- 北京 ：外语教学与研究出版社，2022.12（2023.2重印）

ISBN 978-7-5213-4052-5

Ⅰ. ①北… Ⅱ. ①北… Ⅲ. ①文化名城－保护－研究－北京－汉、英 Ⅳ. ①TU984.21

中国版本图书馆 CIP 数据核字（2022）第 202853 号

地图审图号：京 S（2022）029 号

出 版 人　王　芳
责任编辑　王　琳
责任校对　赵璞玉
封面设计　李　高
版式设计　XXL Studio
出版发行　外语教学与研究出版社
社　　址　北京市西三环北路 19 号（100089）
网　　址　http://www.fltrp.com
印　　刷　北京捷迅佳彩印刷有限公司
开　　本　710×1000　1/16
印　　张　11
版　　次　2022 年 12 月第 1 版　2023 年 2 月第 2 次印刷
书　　号　ISBN 978-7-5213-4052-5
定　　价　59.00 元

购书咨询：（010）88819926　电子邮箱：club@fltrp.com
外研书店：https://waiyants.tmall.com
凡印刷、装订质量问题，请联系我社印制部
联系电话：（010）61207896　电子邮箱：zhijian@fltrp.com
凡侵权、盗版书籍线索，请联系我社法律事务部
举报电话：（010）88817519　电子邮箱：banquan@fltrp.com
物料号：340520001

教育部、国家语委重大文化工程

“中华思想文化术语传播工程”成果

BEIJING

编译团队

编写：北京历史文化名城保护委员会办公室
　　　北京市城市规划设计研究院

翻译：童孝华

审订：韩安德　陆　地　黄希玲

EDITORIAL TEAM

Compilers	Office of Beijing Historic City Conservation Commission, Beijing Municipal Institute of City Planning & Design
Translator	TONG Xiaohua
Reviewers	Harry Anders HANSSON, LU Di, HUANG Xiling

序言

习近平总书记指出，北京历史文化是中华文明源远流长的伟大见证，要更加精心保护好，凸显北京历史文化的整体价值，强化“首都风范、古都风韵、时代风貌”的城市特色。

2022 年是我国设立历史文化名城制度暨北京成为首批国家历史文化名城 40 周年。40 年来，特别是党的十八大以来，北京历史文化名城保护事业取得长足进步，确定了北京历史文化名城保护体系，基本形成了“空间全覆盖、要素全囊括、全过程保护、全社会参与”的保护格局，古都北京在新时代焕发生机、绽放光彩。

为帮助公众更好地了解北京历史文化名城，更好地保护好、传承好、利用好北京丰富的历史文化资源，我们系统梳理北京“历史沿革与营城理念”“保护体系与保护对象”“保护要求与保护措施”三方面的专有名词，从中筛选出 110 个“关键词”。排序时综合考虑《北京历史文化名城保护条例》中相关词条表述顺序及词条重要性等。释义则依据有关国际组织文件、我国法律法规条文，并结合学术研究成果和北京名城保护实践经验。各词条名称及相应释义均适用于北京市域空间范围。

北京是全国政治中心、文化中心、国际交往中心和科技创新中心。新时代首都发展要求我们更好地保护传承好北京的历史文化遗产。展望未来，我们将继续坚持首善标准，全力服务保障首都功能，做好历史文化名城保护工作，擦亮中华文明“金名片”。

感谢“中华思想文化术语传播工程”支持出版中英文对照版的《北京历史文化名城保护关键词》，这将有助于海内外读者了解北京历史文化，传承城市历史文脉，推动文明交流互鉴。

Preface

General Secretary Xi Jinping stated that Beijing's historic and cultural heritage bears exceptional value in witnessing the long history of Chinese civilization. More meticulous measures shall be taken for its safeguarding, so as to highlight its value as a whole and reinforce the city's characteristics and charm as China's current and ancient capital.

The year 2022 marks the 40th anniversary of China's establishment of the system of historic cities and Beijing becoming one of the first of such cities. During the past 40 years, especially since the 18th National Congress of the Communist Party of China (CPC), great progress has been made in the protection of the historic city of Beijing. The city has gradually developed a conservation system characterized by "full spatial coverage, full inclusion of key elements, whole-process protection, and participation by all of society." The ancient capital of Beijing has been revitalized, with its charm blossoming in the new era.

In order to help the public better understand, protect, and utilize the city's rich historic and cultural resources and pass it on to future generations, we have systematically reviewed the concepts from three aspects—history and ideas of urban planning; the conservation system and protected elements; and requirements and measures for conservation, publishing 110 key concepts here. We have listed the concepts in an order based on their place and importance as indicated in the "Regulations on the Conservation of the Historic City of Beijing." We have defined the concepts based on the documents of relevant international organizations, Chinese laws and regulations, as well as the results of academic research and practical experience in the conservation of the city. The names and definitions of the concepts are applicable to the spatial scope of Beijing.

Beijing is the national political center, cultural center, center for international exchanges, and center for scientific discovery and technological innovation. Beijing's development in the new era calls for better safeguarding and transmission of its historic and cultural heritage. In the years to come, we will continue to adhere to the highest standards, do our best to safeguard Beijing's functions as China's capital and ensure the conservation of this historic city and polish this "golden card" of Chinese civilization.

Our thanks go to the Key Concepts in Chinese Thought and Culture-Translation and Communication Project for making this Chinese-English version of *Conservation of the Historic City of Beijing: Key Terms* possible. It is hoped that this book will help readers in China and elsewhere to understand Beijing as a historic city and thus promote the exchanges and mutual learning between Chinese and other civilizations.

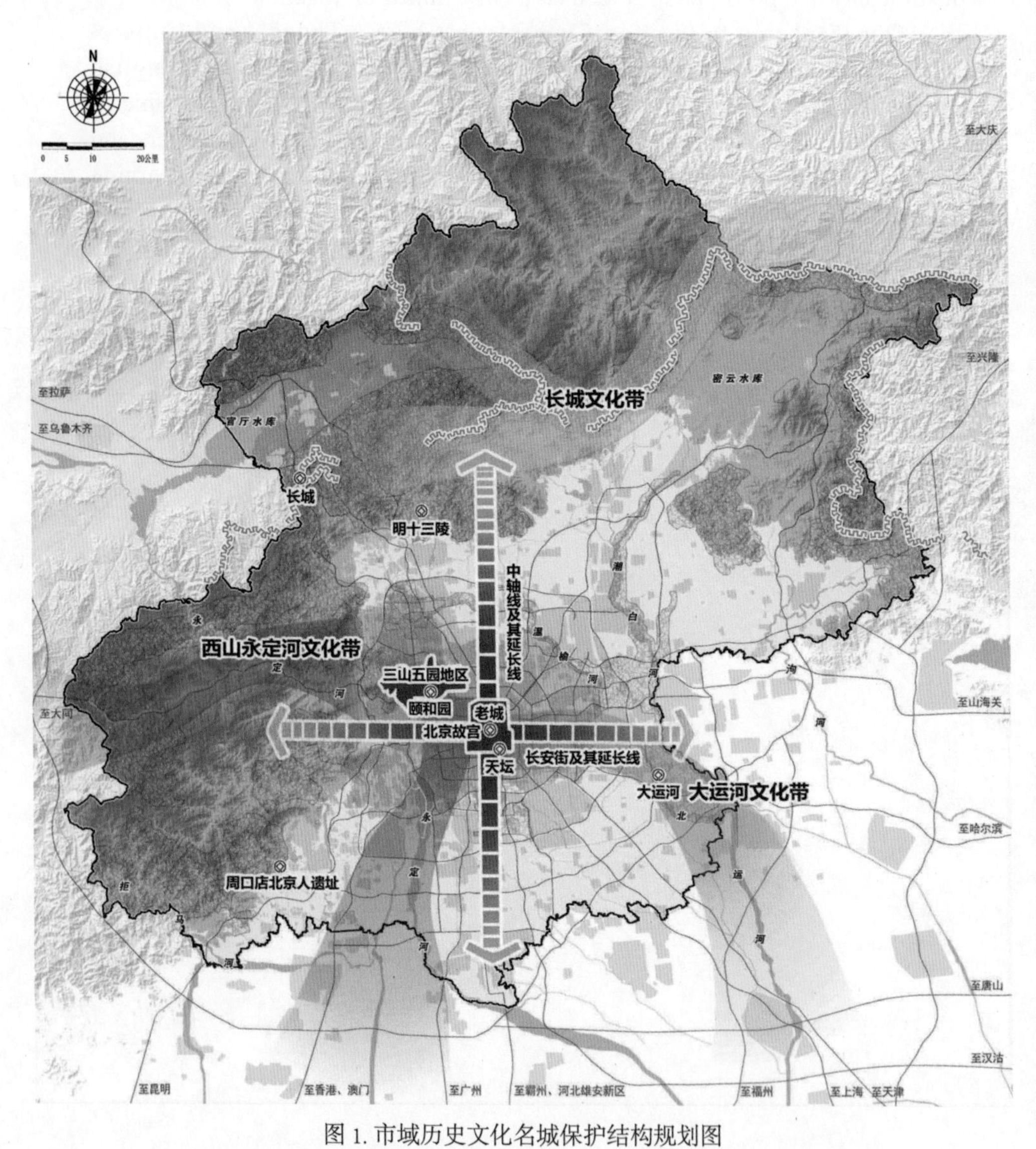

图 1. 市域历史文化名城保护结构规划图

Figure 1. The Conservation Framework for the Historic City of Beijing

图 例

Legend

两轴

Two Axes

重点保护区域

Key Conservation Areas

三条文化带

Three Cultural Belts

世界遗产

World Heritage Sites

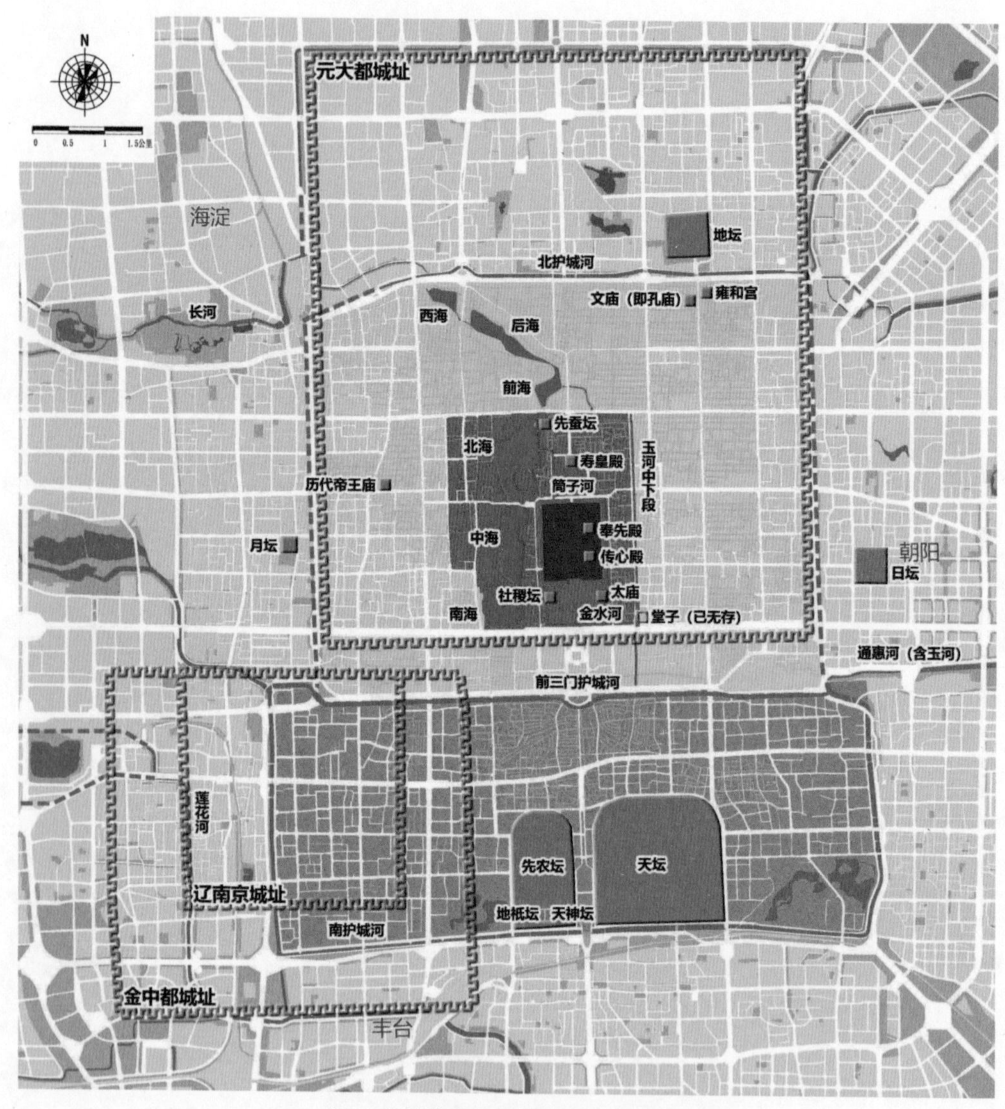

图 2. 老城传统空间格局保护示意图
Figure 2. The Configuration of the Old City of Beijing

图 例

Legend

内城

The Inner City

外城

The Outer City

宫城

The Palace City

皇城

The Imperial City

九坛八庙

Nine Altars and Eight Temples

六海八水

Six Lakes and Eight Rivers

城址遗存

City-Site Remains

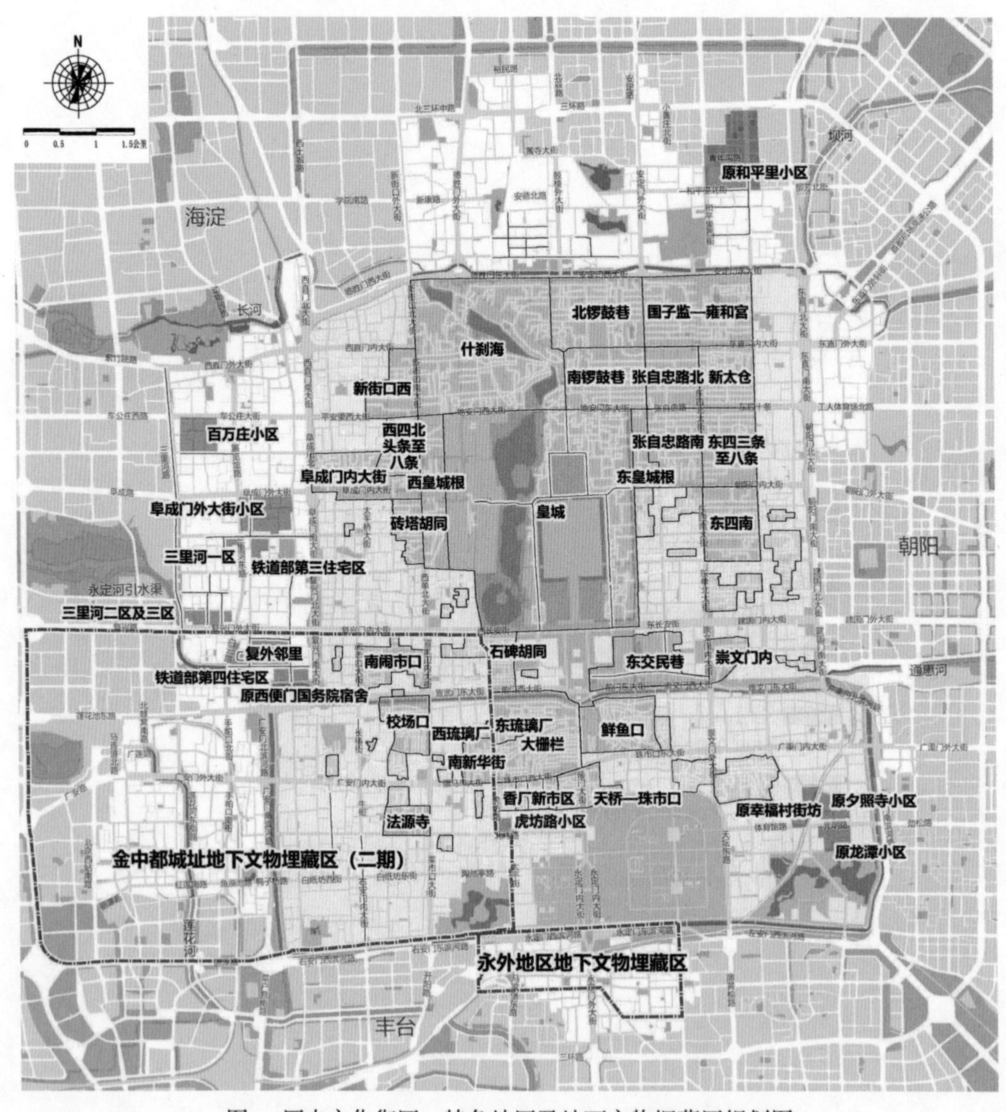

图 3. 历史文化街区、特色地区及地下文物埋藏区规划图
Figure 3. Map of Historic Conservation Areas, Areas of Special Quality and Underground Archaeological Remains

图 例
Legend

历史文化街区
Historic Conservation Areas

特色地区
Areas of Special Quality

其他成片传统平房区
Other Areas of Traditional Contiguous One-Story Houses

地下文物埋藏区
Underground Archaeological Remains

传统胡同
Traditional *Hutongs*

历史街巷
Historic Streets and Alleys

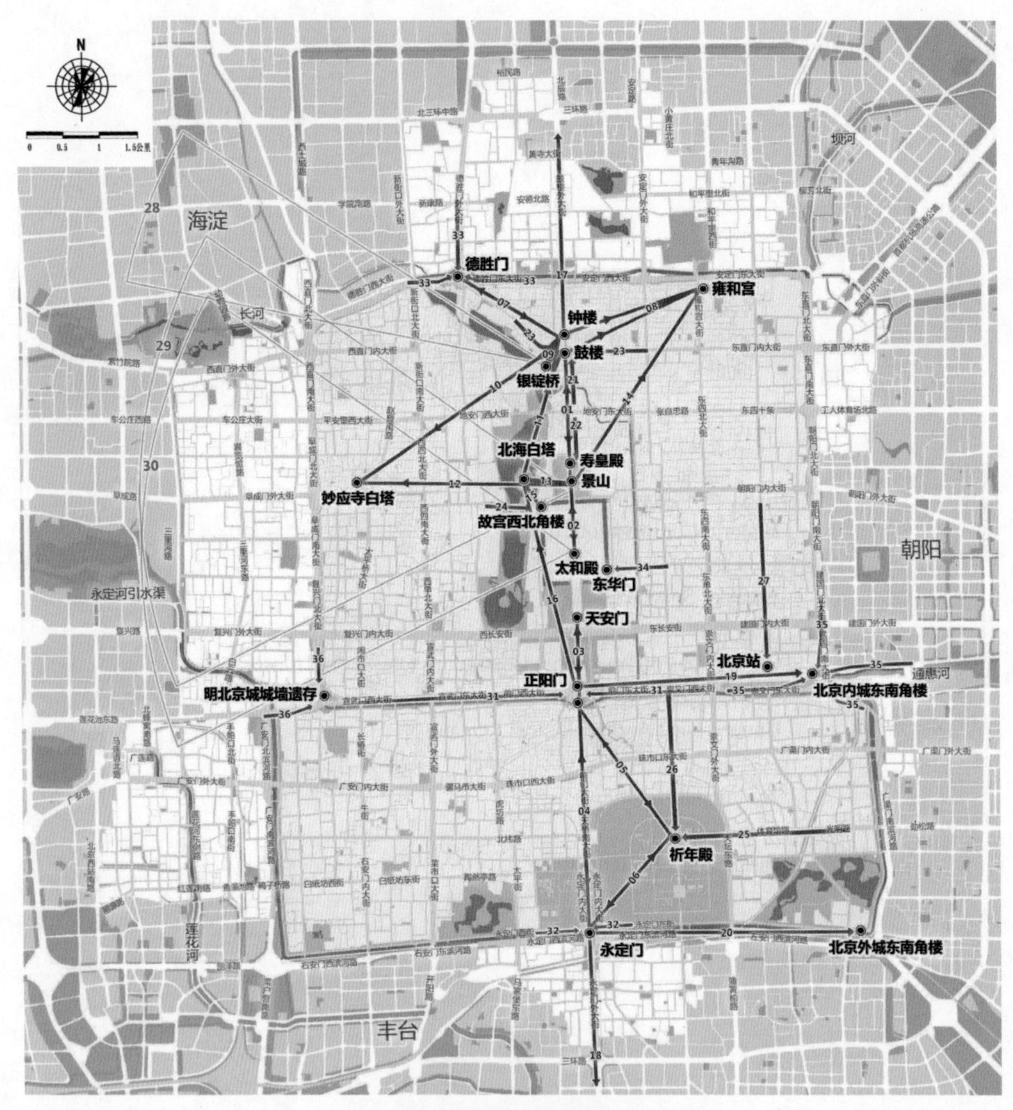

图 4. 景观视廊保护控制规划图
Figure 4. Map of Visual Corridors

图 例
Legend

战略性景观标志物
Strategic View Targets

看历史类景观视廊
Visual Corridors of Historic Sites

看山水类景观视廊
Visual Corridors of Mountains and Waters

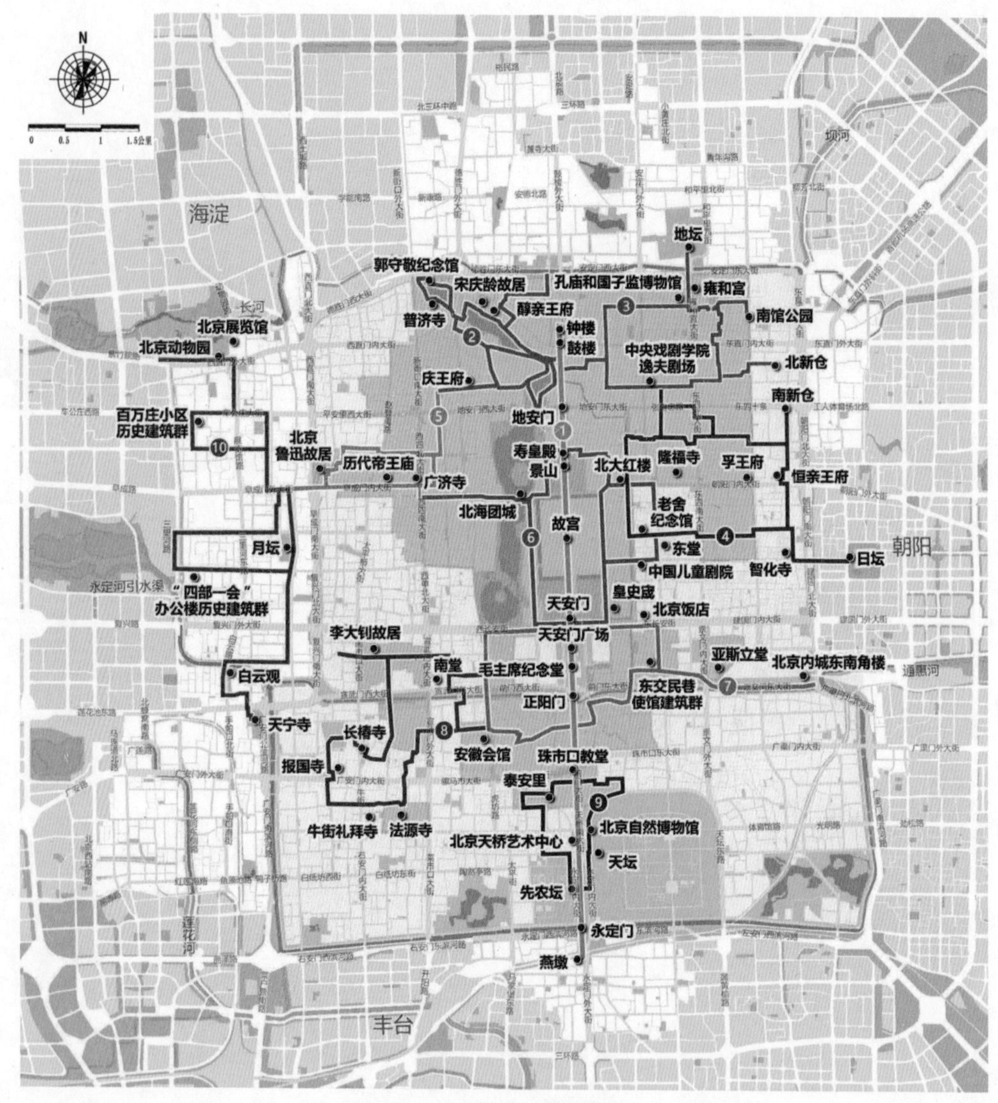

图 5. 文化探访路示意图

Figure 5. Map of Culture Roads

图 例

Legend

历史文化街区
Historic Conservation Areas

文化探访路
Culture Roads

文化精品点
Scenic Spots

目　录
Contents

第二篇　保护体系与保护对象 035

Part II　The Conservation System and Protected Elements

第一篇
历史沿革与营城理念
Part I
History and Ideas of Urban Planning

Běijīng jiànchéng shǐ

北京建城史

History of Beijing as a City

北京有 3,000 多年的建城史。最早可追溯至西周初年分封的蓟、燕两个诸侯国，蓟城和燕都的建立标志着北京建城史的开端。蓟城大约位于今西城区广安门附近，燕都位于今房山区琉璃河镇，以城址为中心的商周时期文化遗存范围超过 20 平方千米。东周初年，燕兼并蓟，并迁都至蓟城，其后历西汉、东汉、魏晋南北朝、隋、唐至五代十国，蓟城作为北方军事重镇的战略地位日益突出。公元 938 年，辽以蓟城为基础建设陪都南京。

Beijing has a history of more than 3,000 years as a city, dating back to the two vassal states of Ji and Yan, which were established as fiefs of the early Western Zhou Dynasty. The construction of the cities of Jicheng (Ji City) and Yandu (Yan Capital) marks the beginning of Beijing's history as a city. Jicheng was located near present-day Guang'anmen in Xicheng District, and Yandu was situated in present-day Liulihe Town in Fangshan District, with the cultural remains of the Shang and Zhou dynasties covering more than 20 square kilometers. In the early years of the Eastern Zhou Dynasty, Yan annexed Ji and moved its capital to Jicheng. Afterwards, through the Western and Eastern Han dynasties, Wei, Jin, the Northern and Southern Dynasties, Sui, Tang, and the Five Dynasties and Ten Kingdoms period, Jicheng became more and more prominent as a military stronghold in the north. In 938, Liao used the old site of Jicheng for constructing its auxiliary capital–Nanjing (the Southern Capital).

Běijīng jiàndū shǐ

北京建都史

History of Beijing as a Capital

北京有 870 年的建都史。金天德三年（公元 1151 年）迁都燕京，天德五年（公元 1153 年）在辽南京城基础上兴建金中都，成为北京正式建都的开始，元明清三朝先后建都于此。1949 年 9 月 27 日，中国人民政治协商会议第一届全体会议一致通过中华人民共和国的首都定于北平，即日起北平改名北京。

Beijing has been a capital city for 870 years. The Jin Dynasty moved its capital to Yanjing (Beijing) in the year 1151. Two years later, it built its capital city of Zhongdu on the site of the Liao Dynasty's southern capital, which marked the beginning of Beijing as the sole official capital. The Yuan, Ming and Qing dynasties all made Beijing their capital. On September 27, 1949, the First Plenary Session of the Chinese People's Political Consultative Conference unanimously approved Beiping as the capital of the People's Republic of China and changed the city's name back to Beijing with immediate effect.

Jīn Zhōngdū

金中都

Zhongdu of the Jin Dynasty

金朝都城。始建于金天德五年（公元 1153 年）。金中都城址位于北京市区西南，其西南城角在今丰台区凤凰嘴街附近，西北城角在今军事博物馆南，东南城角在今北京南站南，东北城角在今北京第三十一中学南。总面积约 22 平方千米。

The capital city of the Jin Dynasty, Zhongdu, was first built in 1153 in the southwest of present-day Beijing. Zhongdu, with a walled area of about 22 square kilometers, reached the vicinity of present-day Fenghuangzui Street in Fengtai District to the southwest, the Military Museum of the Chinese People's Revolution to the northwest, the area south of Beijing South Railway Station to the southeast, and south of present-day Beijing No. 31 Secondary School to the northeast.

Yuán Dàdū

元大都

Dadu of the Yuan Dynasty

元朝都城。始建于元至元四年（公元 1267 年）。元大都城址位于今北京市区中心，北至北土城路，南至长安街，东西至二环路。元大都时期形成的城市轴线、道路格局和河湖水系延续至今。位于建国门的北京古观象台即建在元大都东南角楼基址之上。总面积约 53 平方千米。

As the capital city of the Yuan Dynasty, Dadu was first built in 1267. Dadu, with a walled area of about 53 square kilometers, was located in the center of present-day central Beijing, reaching north as far as Beitucheng Road, south as far as Chang'an Avenue, and to the 2nd Ring Road to the east and west. The city axis, road pattern, and river and water systems formed during the Yuan Dynasty remain as they were to this day. The Beijing Ancient Observatory at Jianguomen gate was built on top of the foundation of the southeast corner tower of Dadu.

Míng-Qīng Běijīngchéng

明清北京城

The Historic Center of Beijing in the Ming and Qing Dynasties

明清两朝的都城。始建于明洪武元年（公元1368年），明嘉靖三十二年（公元1553年）扩建，形成由宫城、皇城、内城、外城组成的四重城垣，整体呈“凸”字形。明清北京城城址位于今北京市区中心，涵盖明清时期北京城护城河及其遗址以内的区域。总面积约63.8平方千米。

Beijing was the capital of the Ming and Qing dynasties. The historic center of Beijing in the Ming and Qing dynasties was first constructed in 1368 under the Emperor Hongwu and expanded in 1553 under the Emperor Jiajing. It consisted of the rectangular Outer City and the square-shaped Inner City which surrounded the Imperial City which in turn had the Palace City at its center. With a total area of about 63.8 square kilometers, it was located in present-day central Beijing, covering the area within the old city moat as it existed in the Ming-Qing period.

nèichéng

内城

The Inner City

明清北京城“凸”字形的北部区域。明洪武、永乐年间在元大都大城基础上改建，历数十年至明正统年间基本成型。明清时期内城中沿皇城四周分布大量居民区、寺庙等，城中商业繁华、民俗活动丰富。内城北至北二环路，南至前三门大街，东西至二环路。总面积约 36 平方千米。

The Inner City was the square-shaped northern part of the City of Beijing in the Ming and Qing dynasties. It was rebuilt on the site of the Yuan Dynasty capital Dadu during the reigns of Ming Emperors Hongwu and Yongle and was basically completed during the reign of Ming Emperor Zhengtong following decades of construction. During the Ming and Qing dynasties, a large number of residential areas and temples were distributed around the imperial city, and the entire city was rich in commerce and folklore activities. Covering a total area of about 36 square kilometers, the Inner City reached the modern North 2nd Ring Road to the north, to Chongwenmen-Qianmen-Xuanwumen Street in the south, and the 2nd Ring Road in the east and west.

wàichéng

外城

The Outer City

明清北京城“凸”字形的南部区域。明嘉靖三十二年（公元 1553 年）为加强安全防御，在北京内城南部增筑外城，将天坛、先农坛纳入。外城北至前三门大街，东、西、南至二环路。总面积约 26.7 平方千米。

The Outer City was the rectangular southern part of the City of Beijing in the Ming and Qing dynasties. This part of the city was added to strengthen the security and defense capabilities of the city and incorporated the Temple of Heaven and the Altar of the God of Agriculture. With a total area of about 26.7 square kilometers, the Outer City covered the area from Chongwenmen-Qianmen-Xuanwumen Street in the north to the 2nd Ring Road in the east, west and south.

gōngchéng

宫城

The Palace City

明清两朝皇家宫廷所在地，即今北京故宫。明永乐初年在元大都宫城基址上兴建，建成于永乐十八年（公元 1420 年），位于皇城的核心部位。宫城涵盖明清时期筒子河及其以内的区域，总面积约 1 平方千米，其中宫城城墙内约 0.72 平方千米。北京故宫是中国现存古代宫城的最高典范，是世界上规模最大、保存最完整的木结构宫殿建筑群。

The Palace City, the present-day Forbidden City in Beijing, was the location of the Ming and Qing dynasty palaces. Located in the core part of the imperial city, it was built based on the palaces in 1420. The Palace City consisted of a walled area with buildings inside, surrounded by a moat (the Tongzi River). All of it covered about 1 square kilometer of which about 0.72 square kilometers were within the walls. The Forbidden City in Beijing is the finest surviving example of an old palace city in China and is the largest and best preserved wooden palace complex in the world.

huángchéng

皇城

The Imperial City

内城中围护宫城的皇家领域，曾有皇城墙环绕。始建于元代，明清两代持续发展演变，以宫城为核心，以传统中轴线为骨架，有序布局皇家宫殿、坛庙建筑群、苑囿、衙署库坊、居住宅院。总面积约 6.8 平方千米。

As part of the Inner City that enclosed the Palace City, the Imperial City was once surrounded by its own walls. With a total area of about 6.8 square kilometers, it was first built in the Yuan Dynasty and continued to be reconstructed and evolve in the Ming and Qing dynasties. With the Palace City at the core and with a traditional central axis running through this framework, the Imperial City was an orderly layout of royal palaces, temple complexes, gardens, government offices, warehouses and workshops, and residential buildings.

sì chóng chéngkuò

四重城廓

Quadruple-Walled City Framework

由明清北京城的宫城、皇城、内城、外城四重城垣的轮廓构成的独特城市格局，是老城整体保护的十个重点之一。

The City of Beijing in the Ming and Qing dynasties had a unique layout in that it comprised the Palace City, the Imperial City, the Inner City, and the Outer City. This quadruple-walled city framework is one of the ten priorities for the overall conservation of the Old City of Beijing.

Běijīng lǎochéng

北京老城

The Old City of Beijing

原称“北京旧城”，大致为北京市区二环路以内的区域（面积略小于明清北京城），总面积约 62.5 平方千米。北京老城是北京历史文化资源最为丰富和集中的区域。《北京城市总体规划（2016 年—2035 年）》将“旧城”改称“老城”，体现了城市发展理念的转变以及对历史积淀的尊重。

The “Old City of Beijing” is about 62.5 square kilometers (slightly smaller than the walled city in Ming and Qing times). This part of the city roughly covers the area within the 2nd Ring Road in downtown Beijing. The Old City is the area with the richest and most concentrated historical and cultural resources in Beijing. The “Old City” was originally referred to as the *jiu cheng* (旧城, “*jiu*” means “derelict”), but the *Beijing Master Plan (2016–2035)* adopted the name *lao cheng* (老城, “*lao*” means “old”), using a word for “old” that often indicates familiarity but also respect. This reflected a new vision for urban development and respect for the historic remains.

Běijīng zhōngzhóuxiàn

北京中轴线

Beijing Central Axis

元明清三朝都城规划和建设的核心，以及北京老城空间布局和功能组织的统领。传统中轴线始建于元代，发展完善于明清，居于北京老城中央、纵贯南北，北端自鼓楼、钟楼，向南经过万宁桥、景山、紫禁城、正阳门，南端至永定门，全长约 7.8 千米。北京中轴线是中国现存规模最为恢宏、保存最为完整的传统都城中轴线，是都城中轴线的典范之作。

The traditional central axis of Beijing was at the core in the planning and construction of the capital city in the Yuan, Ming and Qing dynasties, and a guide in the spatial layout and functional organization of the Old City of Beijing. Taking its form in the Yuan Dynasty and enlarged and improved in the Ming and Qing dynasties, the central axis runs north and south through the center of the old city. It stretches about 7.8 kilometers from the Bell and Drum Towers in the north to Yongdingmen gate in the south via the Wanning Bridge, the artificial hill Jingshan, the Forbidden City, and the Zhengyangmen gate. It is the most magnificent, well preserved and exemplary traditional central axis of China's capital cities.

yǐ zhōng wéi zūn

以中为尊

Center Being the Place of Dignity

中国古人在观象授时实践中形成的理念。汉字“中”所象之形表示了辨方正位定时的方法，通过此种方法提供农耕时代最为重要的观象授时服务，是君权产生的基础，并衍生出“中庸”“中和”“中正”“允执厥中”“居中而治”等理念。以中为尊，彰显了古代时空观，是中国古代都城营建、城市规划和建筑布局的重要原则。

The ancient Chinese came to see the center as the place of dignity when they observed the sky to determine the time. The Chinese character for “center” (中) represents the method of “determining the four sides of the capital and its principal sites.” This method made it possible to provide the most important service in the agricultural age by observing the sky to determine and tell the seasons. It was the basis of the legitimacy of the monarchy. Other concepts of traditional Chinese thought were also derived from this practice, such as the “mean,” “balanced harmony,” “exactitude,” “concentration of mind is required for sticking to the path of justice and uprightness” and “governing at the center.” According dignity to the center is thus an ancient element in the Chinese view of space and time and it was an important principle in capital city planning and construction, as well as architectural layout.

jiǔjīng jiǔwěi

九经九纬

Nine North-South Roads and Nine East-West Roads

出自《周礼·考工记》。指在城市中建设九条南北大道和九条东西大道，经纬交错形成棋盘路网。九经九纬，是中国传统理想都城的营建法则之一。

First appearing in the old text *Kaogong Ji* (*The Artificers' Record*) of *The Rites of Zhou*, this concept refers to the construction of nine north-south roads and nine east-west roads in the capital city to form an orthogonal grid street layout. This is a construction principle that represents a traditional ideal for capital cities in China.

zuǒ zǔ yòu shè

左祖右社

A Temple on the Left (East) Side and an Altar on the Right (West) Side

出自《周礼·考工记》。指以朝南为正向，在宫城左（东）侧设太庙以祭祀祖先，在宫城右（西）侧设社稷坛以祭祀土地神、五谷神。左祖右社，是中国传统理想都城的营建法则之一。

First appearing in the early text *Kaogong Ji* (*The Artificers' Record*) of *The Rites of Zhou*, this concept prescribes that a south-facing temple for ancestor worship must be built east of the palace city and an altar for the worship of the gods of land and grain must be built on the west side. It is one of the construction principles for traditional ideal capital cities in China.

miàn cháo hòushì

面朝后市

Palace in the Front, Market in the Back

出自《周礼·考工记》。指以朝南为正向，皇帝处理朝政的宫殿在前（南），百姓买卖商品的市场在后（北）。面朝后市，是中国传统理想都城的营建法则之一。

First appearing in the old text of *Kaogong Ji* (*The Artificers' Record*) of *The Rites of Zhou*, this is the notion that the palace where the Emperor conducted affairs of state was in front and facing south, while the marketplace where the people bought and sold goods was behind, north of the palace. Having a market in the back is one of the construction principles for traditional ideal capital cities in China.

jiǔ tán bā miào

九坛八庙

Nine Altars and Eight Temples

明清时期皇家祭祀建筑的代称。“九坛”一般指天坛、地坛、日坛、月坛、先农坛、先蚕坛、天神坛、地祇坛和社稷坛；“八庙”一般指太庙、文庙（即孔庙）、历代帝王庙、奉先殿、寿皇殿、雍和宫、传心殿、堂子（已无存）。“国之大事，在祀与戎。”（《左传·成公十三年》）祭祀在中国传统礼制中占据着重要地位，各类坛庙是都城不可或缺的建筑类型。

This term denotes the imperial ritual buildings of the Ming and Qing dynasties. The nine altars generally refer to the Temple of Heaven, Altar of the Earth, Altar of the Sun, Altar of the Moon, Altar of the God of Agriculture, Altar of the Goddess of Silkworms, Altar of the Spirits of Heaven, Altar of the Spirits of the Earth, and Altar of Land and Grains. The eight temples generally refer to the Imperial Ancestral Temple, Temple of Confucius, Temple of Past Monarchs, Hall for Ancestral Worship, Hall of Imperial Longevity, Yonghe Palace, aka "the Lama Temple," Hall of the Transmission of the Mind, and the Manchu Shamanist Tangzi Temple (which no longer exists). As *Zuo Zhuan* says, "For a state the great events are the sacrifices and warfare." Sacrifices occupied an important place in the traditional Chinese ritual system, and various types of altars and temples were indispensable types of buildings in capital cities.

liù hǎi bā shuǐ

六海八水

Six Lakes and Eight Rivers

北京历史河湖水系与水文化遗产的代表。“六海”指北海、中海、南海、西海、后海、前海；“八水”指通惠河（含玉河）、北护城河、南护城河、筒子河、金水河、前三门护城河、长河、莲花河。六海八水，是中国古代山水文化在北京城市规划中的体现。

Beijing's historic lakes and watercourses are part of the city's cultural heritage. The "six lakes" are the Beihai (North Lake), Zhonghai (Central Lake), Nanhai (South Lake), Xihai (West Lake), Houhai (Back Lake), and Qianhai (Front Lake). The "eight rivers" are the Tonghui River (including the Jade River), the North Moat, the South Moat, the Tongzi River, the Jinshui River, the Chongwenmen-Qianmen-Xuanwumen Moat, the Changhe River, and the Lianhua River. The six lakes and eight rivers were the embodiment of an ancient and idealized Chinese urban landscape culture in Beijing's urban planning.

qípán lùwǎng

棋盘路网

Orthogonal Grid Street Layout

北京老城呈横平竖直、棋盘式布局的主干道路网。棋盘路网是北京老城道路网的基本特征，也是老城整体保护的十个重点之一。棋盘路网主要由多条东西向道路和南北向道路组成，东西向道路主要包括平安大街、朝阜路、长安街、前三门大街、两广大街；南北向道路主要包括朝阳门北小街一线、东单北大街一线、王府井大街一线、西单北大街一线、赵登禹大街一线。棋盘路网，是中国传统营城理念在北京老城道路建设中的重要体现与传承发展。

The Old City of Beijing has a network of horizontal and vertical main streets laid out like a Chinese chessboard. This is the basic feature of the street system and is one of the ten priorities in the overall conservation of the Old City of Beijing. The chessboard layout of streets mainly consists of several east-west streets and north-south streets. The main east-west ones are Ping'an Street, Chaofu Road, Chang'an Avenue, Chongwenmen-Qianmen-Xuanwumen Street and Guang'anmen-Guangqumen Street; the main north-south ones are Chaoyangmen North Street, Dongdan North Street, Wangfujing Street, Xidan North Street and Zhao Dengyu Street. The chessboard road network is an important embodiment, continuation and development of the traditional Chinese concept of urban planning in Beijing's street layout.

chuántǒng fēngmào

传统风貌

Traditional Features

在一定区域内所反映历史文化特征的城乡景观和自然人文环境的整体面貌。北京老城是集中展现明清北京传统风貌的重要区域，拥有气势恢宏、庄重威严的宫殿寺庙，青砖灰瓦、中式坡顶的典型民居建筑，窄小幽静、肌理清晰的胡同街巷，平缓开阔、壮美有序的整体空间形态，红黄金碧、灰院素城的传统色彩意象，整体展现出中国传统文化的独特风韵与深邃内涵。

The traditional feature is the overall appearance of the urban and rural townscape and natural and cultural environment that reflect the historical and cultural characteristics in a certain area. The Old City of Beijing is an important area that displays the traditional townscape of Beijing in the Ming and Qing dynasties. The city has magnificent and majestic palaces and temples, typical residential buildings with gray bricks and tiles and Chinese sloped rooftops, narrow, quiet and well laid out streets and alleys (the latter known as *hutongs*). It is a gentle, open and magnificently ordered spatial layout with a traditional color pattern of red, yellow, gold, green and gray. The Old City displays the unique charm and profound connotations of the Chinese traditional culture.

jiēxiàng jīlǐ

街巷肌理

Street and Alley Fabrics

各历史时期形成的街巷、胡同以及沿线建筑物、构筑物所共同构成的能够反映传统风貌的空间格局与结构。如北京南锣鼓巷历史文化街区的鱼骨式街巷肌理、大栅栏历史文化街区的斜街式街巷肌理等。

The city's spatial fabrics, that is to say the streets, alleys and *hutongs* as well as the buildings and structures along them, reflect the traditional styles of different historical periods. For example, Beijing has the fishbone street fabrics of the Nanluoguxiang Historic Conservation Area and the fabrics of diagonal street and alleys of the Dashilar Historic Conservation Area.

chuántǒng jiànzhù sècǎi

传统建筑色彩

Traditional Architectural Colors

由青（蓝）、赤（红）、黄、白、黑组成的五色体系。中国传统建筑色彩广泛运用于北京的传统建筑，并基于传统礼制和传统文化形成了明确的色彩使用规则。明清时期，皇宫及敕建寺庙多采用红墙、黄瓦、白色石材，王府宅邸多采用红漆柱、绿瓦，民居则采用黑门、灰瓦。

The traditional colors on buildings are blue, red, yellow, white and black. This five-color system is widely used on Beijing's traditional-style buildings and strict rules have been established for color use based on time-honored rituals and culture. During the Ming and Qing dynasties, the imperial palaces and imperial temples mostly had red walls, yellow tiles and white stonework, the princes' residences mostly had red-painted columns and green tiles, and the common people's residences had black doors and gray tiles.

hútòng sìhéyuàn chuántǒng jiànzhù xíngtài

胡同—四合院传统建筑形态

The Traditional Architectural Form of *Hutongs* and *Siheyuan* Compounds

胡同与四合院相互依存构成的老城传统空间形态基底，是老城整体保护的十个重点之一。胡同是公共空间，是连接四合院的网络；四合院是私人空间，是构成胡同界面和风貌的要素。

Traditionally interdependent, the *hutongs* (alleys) and *siheyuan* residential courtyard compounds are the basis of traditional urban spatial layout and one of the ten priorities of the overall conservation of the Old City of Beijing. The *hutongs* are public spaces that connect the network of courtyard compounds; the courtyard houses are private spaces that constitute the interface and the streetscape of the *hutongs*.

Běijīng sìhéyuàn

北京四合院

The *Siheyuan* Compounds of Beijing

北京传统民居的总称。通常是由东、西、南、北四面房屋围合成一个院落单元，标准的一进院由以中轴对称的正房（北房）、东西厢房、倒座房（南房）围合而成，多为青砖灰瓦、砖木结构、中式坡顶的单层建筑。在此基础上可纵向发展，形成二进院、三进院等；也可横向发展，形成二路或三路并列的院落。北京四合院是中国传统民居的优秀代表，是传统居住观念与社会生活结构的集中体现。

Siheyuan is the general name for Beijing's traditional residential buildings. A *siheyuan* compound is usually a courtyard surrounded by houses on all four sides. Such a standard single-story unit, mostly with gray bricks and tiles, a framework of bricks and timbers, and a traditional Chinese sloping roof, is normally made up of a south-facing main building on the northern side, adjoining side buildings to the east and west of the courtyard, and a north-facing house on the southern side. Such a compound can be enlarged upward to form a two or three sets of buildings or horizontally to form two or three rows. These Beijing residential compounds are excellent examples of traditional housing and a concentrated embodiment of social structure.

jǐngguān shìláng

景观视廊

Visual Corridors

城市中特定观测点与特定建筑或者山林景观等目标物之间形成的视线廊道。《首都功能核心区控制性详细规划（街区层面）（2018 年—2035 年)》中明确提出 36 条在城市整体层面具有重要标识意义的战略级景观视廊，是老城整体保护的十个重点之一。景观视廊体现了营城美学。如北京的鼓楼望景山万春亭、永定门城楼望正阳门城楼等。

Visual corridors are formed between observation points in a city and landmarks such as specific buildings or hills and trees. The "Regulatory Plan for the Core Area of the Capital (Block Level) (2018–2035)" specifies 36 strategic visual corridors that are of overall importance for the city and makes these corridors one of ten priorities for the conservation of Beijing's Old City. These visual corridors reflect the aesthetics in urban planning. For example, Beijing's Drum Tower offers a good view of the Wanchun Pavilion in Jingshan Park, and the Yongdingmen gate tower provides a good view of the Zhengyangmen gate tower.

jiēdào duìjǐng

街道对景

Street View Corridors

从城市街道眺望街道尽头特定建筑形成的视线廊道。如北京的前门大街望箭楼、光明路望天坛祈年殿、永定门内大街望永定门城楼、地安门大街望鼓楼等。

Street view corridors are what one sees when looking along a street toward specific buildings at the far end. For example, in Beijing, Qianmen Street offers a good view of the Arrow Tower of Zhengyangmen gate, Guangming Road offers a good view of the Hall of Prayer for Good Harvests at the Temple of Heaven, Yongdingmennei Street offers a good view of the Yongdingmen gate tower, and Di'anmen Street provides a good view of the Drum Tower to its north.

chéngshì dì-wǔ lìmiàn

城市第五立面

The Fifth Façade of the City

从空中俯瞰城市上部空间的整体意象，由建筑屋顶、街巷道路、开敞空间、自然风貌、河湖水系和绿化植被等要素构成。建筑屋顶是构成城市第五立面的核心要素。

If we have a wide view of a city from the air, we see a “fifth façade” which is composed of roofs, streets and roads, open spaces, parkland, river and lake system, and green vegetation. Roofs on buildings are the core element of the fifth façade.

sān shān wǔ yuán

三山五园

Three Hills and Five Gardens

始建于辽金、形成于清鼎盛时期，以北京西北郊皇家园林为代表的各历史时期文化遗产的统称。三山五园是清代帝王休憩、游赏、居住、理政之所，“三山”指香山、玉泉山和万寿山，“五园”指静宜园、静明园、颐和园、圆明园和畅春园。三山五园地区的园林规划与设计构思独特，巧妙融合山体、水系，集中反映了中国天人合一的哲学理念，也是世界造园史上的杰作。

This is a collective term for the cultural heritage represented by the imperial gardens in Beijing's northwestern suburbs. The "three hills" are Xiangshan (Fragrant Hill), Yuquanshan (Jade Spring Hill) and Wanshoushan (Longevity Hill), and the "five gardens" are the Jingyiyuan (Garden of Tranquility and Pleasure), Jingmingyuan (Garden of Tranquility and Brightness), Yuanmingyuan (Garden of Perfection and Brightness) and Changchunyuan (Garden of Everlasting Spring) as well as the Summer Palace. They were first laid out and built in the Liao and Jin dynasties and completed at the height of the Qing Dynasty. They were the places where the Qing emperors rested, enjoyed the scenery, lived and worked. The gardens in this area were uniquely planned and designed to be clearly integrated with the hills and waters. They reflect the Chinese philosophy of humans living in harmony with nature and are considered as masterpieces in the history of world gardening.

gǔdiǎn yuánlín

古典园林

Classical Gardens

运用工程技术和艺术手段，通过筑山、叠石、理水等手段改造地形、种植树木花草、营造建筑和布置园路等途径创作而成的自然环境和休憩境域，具有师法自然、浑然天成的审美意趣。北京的古典园林包括皇家园林和私家园林，现有遗存以前者为主。皇家园林是皇家生活环境的重要组成部分，整体恢宏富丽，兼具自然野趣，是中国古典园林艺术顶峰时期的代表作和东方造园艺术的杰出范例。如颐和园、圆明园、景山、北海等。

Classical gardens are the natural environment and resting places created by using engineering and artistic means of building hills, stacking rocks and managing water in order to transform the terrain, plant trees and flowers, create buildings and lay out garden paths. These gardens are aesthetically natural and add to the harmony between humans and nature. Beijing's classical gardens include both imperial and private gardens, with the surviving ones mainly belonging to the former category. As an important part of the imperial family's living environment, the gardens were magnificent and rich in their entirety, as well as natural and untamed. They are masterpieces of the peak period of the Chinese classical art of gardening and outstanding examples of East Asian horticulture. These gardens include the Summer Palace and Yuanmingyuan, Jingshan and Beihai parks.

Jīng xī dàotián jǐngguān

京西稻田景观

The Rice Field Landscape in Western Beijing

三山五园地区“山、水、田、园”景观体系的有机组成部分。北京有着悠久的种植稻米的传统，京西稻品种形成于清代。京西稻田穿插于皇家园林之间，形成独具东方意境的田园风光。

The northwestern parts of Beijing with their hills, water, fields, and gardens constitute a “rice-field landscape.” Beijing has a long history of rice cultivation, and the Jingxi Rice (western Beijing rice) variety was developed in the Qing Dynasty. Rice cultivation has been introduced into the former imperial gardens, creating a unique East Asian landscape.

gǔdào

古道

Old Roads

承载某种特定交通运输功能，具有政治、经济、军事和文化等方面价值的历史通道。古道包括商旅古道、军事古道和进香古道等，沿途多分布有古村落、古寺庙等历史遗存。位于北京市门头沟的京西古道，历经千年始终作为北京连接周边地区的通道，担负着内外交通、物产交换、宗教活动和军事防御等功能。

Old roads are historic passage ways that fulfill a specific transportation function and have political, economic, military and cultural values. Serving commercial, military, pilgrimage and other purposes, such roads are mostly lined with age-old villages, temples and other historical remnants. Located in Mentougou District outside the main urban area, the Old West Road has always served as a channel connecting Beijing to the adjacent areas for a millennium, sustaining such functions as inbound and outbound transportation, exchange of goods, religious activities and military defense.

Yānjīng bā jǐng

燕京八景

Eight Sights of Yanjing

北京地区始于金代并持续演变的八处景观单元，命名颇具诗意，是体现中国山水文化和景观设计传统的“题名景观”。史料中多见清乾隆立碑钦定的燕京八景，即：太液秋风、琼岛春阴、金台夕照、蓟门烟树、西山晴雪、玉泉趵突、卢沟晓月、居庸叠翠。在民间还流传着“燕京小八景”：南囿秋风、东郊时雨、银锭观山、西便群羊、燕社鸣秋、长安观塔、回光返照、西直折柳。其中以“银锭观山”最为著名。

The so-called Eight Sights of Yanjing (i.e. Beijing) is a notion that originated in the Jin Dynasty and continued to evolve thereafter. Poetically named, they are “named sights” that reflect Chinese landscape culture and design traditions. In historical texts, the Eight Sights mostly refer to places designated by Emperor Qianlong of the Qing Dynasty, namely Autumn Winds on Taiye Lake, Jade Islet in Shady Springtime, Evening Glow on Golden Terrace, Trees in Mist near the Jimen Gate, Clearing After Snowfall in the Western Hills, Leaping Rainbow over Jade Spring, Moon at Dawn by Lugou Bridge, and The Green Ranges at Juyong Pass. There is also a popular version known as the Eight Little Views of Yanjing, namely, Autumn Winds in Nanyuan, Seasonable Rain in the Eastern Suburbs on the Terrace, Mountain View at Yinding Bridge, The White Sheep at Xibianmen Gate, Autumn Swallows on the Terrace, The Two Pagodas at Chang’an Avenue, Sunlight Reflected at Erlang’s Temple, and Xizhimen Willow Twigs as Farewell Gifts. Among these, the Mountain View at Yinding Bridge is the most famous.

第二篇
保护体系与保护对象
Part II
The Conservation System and Protected Elements

lìshǐ wénhuà míngchéng

历史文化名城

Historic Cities

保存文物特别丰富并且具有重大历史价值或者革命纪念意义，并由国务院核定批准的城市。北京于 1982 年被国务院公布为第一批国家历史文化名城。根据《北京历史文化名城保护条例》（2021 年），北京历史文化名城的范围涵盖本市全部行政区域，总面积约 16,410 平方千米。

Historic cities are those that are particularly rich in historic monuments, are of great historic value or significance for the Chinese revolution, and have been designated and approved as such by the State Council. Beijing was designated in 1982 as one of the first batch of such national historic cities. According to the "Regulations on the Conservation of the Historic City of Beijing" (2021), the conservation area covers the entire administrative area of Beijing, or approximately 16,410 square kilometers in total.

chéngxiāng lìshǐ wénhuà bǎohù chuánchéng tǐxì

城乡历史文化保护传承体系

The Conservation and Inheritance System of Historic and Cultural Heritage in Urban and Rural Areas

以具有保护意义、承载不同历史时期文化价值的城市和村镇等复合型、活态遗产为主体和依托，保护对象主要包括历史文化名城、名镇、名村(传统村落)、街区和不可移动文物、历史建筑、历史地段，与工业遗产、农业文化遗产、灌溉工程遗产、非物质文化遗产、地名文化遗产等保护传承共同构成的有机整体。2021 年 8 月，中共中央办公厅、国务院办公厅印发《关于在城乡建设中加强历史文化保护传承的意见》，首次提出构建城乡历史文化保护传承体系。

This is a framework for preserving the cultural heritage of different periods as well as other complex and living heritage. It is mainly concerned with historic cities, historic towns and villages (traditional villages), historic conservation areas and immovable cultural relics, historic buildings and areas, as well as industrial heritage, agricultural heritage, old irrigation works, intangible cultural heritage, and old place names. This framework for conservation was first put forward by the General Office of the CPC Central Committee and the General Office of the State Council in August 2021 in their "Guidelines on Preserving and Carrying Forward Historic and Cultural Heritage in Urban and Rural Construction."

kōngjiān quán fùgài, yàosù quán nángkuò

空间全覆盖、要素全囊括

Full Spatial Coverage and Inclusion of All Elements

在我国的城乡历史文化保护传承工作中，既要保护单体建筑，也要保护街巷街区、城镇格局，还要保护好历史地段、自然景观、人文环境和非物质文化遗产。《北京历史文化名城保护条例》（2021 年）实现了对历史文化名城保护对象的空间全覆盖、要素全囊括。

In preserving and passing on China's urban and rural historic and cultural heritage, it is necessary to protect not only individual buildings, but also streetscapes and urban districts, as well as historic sites, natural and cultural environments, and intangible cultural heritage. The "Regulations on the Conservation of the Historic City of Beijing" (2021) have achieved full spatial coverage and inclusion of all elements to be protected in a historic city.

Běijīng lìshǐ wénhuà míngchéng bǎohù tǐxì

北京历史文化名城保护体系

The Conservation Framework for the Historic City of Beijing

由四个空间圈层、两大重点区域、三条文化带和九个主要方面构成，简称“四二三九”。“四”指以老城、中心城区、市域和京津冀四个空间圈层搭建名城保护体系；“二”是将北京老城、三山五园地区作为名城保护的两大重点区域；“三”是将长城、大运河、西山永定河三条文化带作为连线成片保护并带动区域更新的重要抓手；“九”指世界遗产、文物、历史建筑、历史文化街区等九方面的保护对象。

Known as “the 4239 Framework” in short, the conservation framework for Beijing consists of four zones, two key areas, three cultural belts and nine major aspects. The four zones refer to the Old City of Beijing, the surrounding central urban area (the six central districts), Beijing Municipality, and the entire Beijing-Tianjin-Hebei region. The two key areas are the Old City of Beijing and the Three Hills and Five Gardens Area. The three cultural belts are the Great Wall Cultural Belt, the Grand Canal Cultural Belt, and the Western Hills and Yongding River Cultural Belt, which will be protected as a contiguous area to drive regional revitalization. The nine major aspects refer to conservation elements such as world heritage, historic monuments, historic buildings and historic conservation areas.

zhòngdiǎn bǎohù qūyù

重点保护区域

Key Conservation Area

北京历史文化名城保护的两大重点区域，即老城和三山五园地区。两个区域直线距离十余千米，历史上通过水陆御道相连，是功能上相辅相成、规模上相近相当、布局上相对相连的两大首都功能重要承载地。

The two most important areas for the conservation of the historic city of Beijing are the Old City and the Three Hills and Five Gardens Area. Slightly over 10 kilometers apart, they were relatively well connected to one another by an imperial route by water and land. These two areas were complementary, similar in scale, and important locations for the function of the capital.

lǎochéng bǎohù shí gè zhòngdiǎn

老城保护十个重点

Ten Priorities for the Conservation of the Old City of Beijing

老城整体保护特别是空间格局和传统风貌保护十个方面的任务。具体包括：(1) 保护北京中轴线；(2) 保护明清北京城“凸”字形城廓；(3) 整体保护明清皇城；(4) 恢复历史河湖水系；(5) 保护老城原有棋盘式道路网骨架和街巷胡同格局，保护传统地名；(6) 保护北京特有的胡同—四合院传统建筑形态，老城内不再拆除胡同和四合院；(7) 分区域严格控制建筑高度，保持老城平缓开阔的空间形态；(8) 保护重要景观视廊和街道对景；(9) 保护老城传统建筑色彩和形态特征；(10) 保护古树名木及大树。

The ten key tasks in the comprehensive conservation of the Old City, especially its spatial pattern and traditional townscape, are the following: (1) preserving the Beijing central axis; (2) preserving the Ming and Qing-dynasty city layout with the walled square-shaped Inner City immediately north of the walled rectangular Outer City; (3) pursuing the overall conservation of the Ming-Qing Imperial City; (4) restoring the historic river and lake systems; (5) preserving the original orthogonal grid street layout and the pattern of alleyways and *hutongs*, and also keeping the traditional place names; (6) preserving the traditional built environment of *hutongs* and *siheyuan* compound compound unique to Beijing, and ensuring that no more such structures are demolished; (7) strictly controlling the building heights by district and maintaining the flat and open space of the Old City; (8) preserving important visual corridors and street view corridors; (9) preserving traditional architectural colors and building shapes; and (10) protecting old and valuable trees as well as large trees.

liǎng zhóu

两轴

Two Axes

东西向的长安街及其延长线、南北向的中轴线及其延长线。两轴垂直相交，是北京城市空间结构的重要统领。长安街始建于元代、定名于明代，在不同历史时期空间范围有所不同。中华人民共和国成立后，长安街伴随首都发展建设而拓展，其中复兴门与建国门之间长约7千米，向东延伸至北京城市副中心和北运河、潮白河，向西延伸至首钢地区、永定河水系、西山山脉，全长约63千米。中轴线始建于元代，在不同历史时期均有发展。中华人民共和国成立后，中轴线伴随北京主办亚运会和奥运会，以及建设北京大兴国际机场等向外延伸，其中钟鼓楼与永定门之间长约7.8千米，向北延伸至燕山山脉，向南延伸至北京大兴国际机场、永定河，全长80多千米。

The two axes are the east-west Chang'an Avenue and its extensions, and the north-south central axis and its extensions. Intersecting at right angles, the two axes are crucial in determining Beijing's urban spatial structure. Chang'an Avenue was first laid out in the Yuan Dynasty and named in the Ming Dynasty, and its spatial extent varied in different historical periods. After the founding of the People's Republic of China, it was expanded along with the development and construction of the capital. The central part of the avenue stretches about seven kilometers between Fuxingmen and Jianguomen gates. Eastward it extends as far as the Beijing Municipal Administrative Center, the north section of the Beijing-Hangzhou Grand Canal and the Chaobai River, and westward it reaches the new Shougang area, the Yongding River area, and the Western Hills range. The street's full length is 63 kilometers. The north-south axis was first built in the Yuan Dynasty and has been developed in different historical periods. After the founding of the People's Republic of China, the central axis was extended along with Beijing's hosting of the Asian Games and the Olympic

Games, the construction of Beijing Daxing International Airport, etc. Totaling more than 80 kilometers in length, about 7.8 kilometers of this axis are in central Beijing from the Bell and Drum Towers in the north to Yongdingmen gate in the south. It extends further north to the Yanshan range and further south to Beijing Daxing International Airport and the Yongding River.

sān tiáo wénhuàdài

三条文化带

Three Cultural Belts

长城文化带、大运河文化带、西山永定河文化带。长城、大运河、永定河在北京境内分别长约 520 千米、80 千米、170 千米。三条文化带以长城、大运河、西山和永定河为主线，串联沿线古村镇、古建筑、古道等各类保护对象，是实现北京历史文化遗产连线、成片整体保护，传承京津冀地区同根同源的历史文脉的重要载体。

The three cultural belts are the Great Wall, Beijing-Hangzhou Grand Canal, and the Western Hills and Yongding River belts. Within the limits of Beijing Municipality, the Great Wall, the Grand Canal and the Yongding River are about 520 kilometers, 80 kilometers and 170 kilometers in length respectively. Along these belts are ancient villages and towns, old buildings and roads, etc. The belts are important carriers of the integrated conservation of Beijing's historic and cultural heritage as well as proof of the shared historic and cultural roots in the Beijing-Tianjin-Hebei area.

lìshǐ wénhuà jiēqū

历史文化街区

Historic Conservation Areas

保留有一定数量的不可移动文物、历史建筑、传统风貌建筑以及传统胡同、历史街巷等历史环境要素，能够较完整、真实地体现历史格局和传统风貌，并具有一定规模的区域。截至 2022 年 4 月，北京已划定 49 片历史文化街区，其中，皇城、大栅栏、东四三条至八条历史文化街区被列入中国历史文化街区。

Historic conservation areas are areas of adequate size that preserve an adequate amount of immovable cultural relics, historic buildings, traditional-style buildings, together with enviroment elements like traditional *hutongs* and historic alleys, etc. These areas are representative of the city's historical layout and traditional features with integrity and authenticity. As of April 2022, Beijing had designated 49 historic conservation. Among the 49 historic conservation areas, the Imperial City, Dashilar, Dongsi Santiao to Batiao are listed national-level historic conservation areas.

shìjiè yíchǎn

世界遗产

World Heritage Sites

根据《保护世界文化和自然遗产公约》和《实施〈世界遗产公约〉操作指南》，符合突出普遍价值标准，满足完整性和（或）真实性条件，具有良好的保护管理状况，由联合国教科文组织世界遗产委员会认定的具有世界意义的文化和（或）自然遗产。截至 2021 年 7 月，北京是全球范围内世界遗产数量最多的城市，有长城、北京故宫、周口店北京人遗址、颐和园、天坛、明十三陵和大运河等 7 处世界遗产。

World heritage sites are cultural and/or natural sites of global significance that are considered by the UNESCO World Heritage Committee as having outstanding universal value, satisfying conditions of integrity and/or authenticity, and having a sound conservation and management system in accordance with the "Convention Concerning the Protection of the World Cultural and Natural Heritage" and its Operational Guidelines. As of July 2021, Beijing is the city with the largest number of world heritage sites worldwide, with seven world heritage sites, namely the Great Wall, the Forbidden City, the Peking Man Site at Zhoukoudian, the Summer Palace, the Temple of Heaven, the Ming Tombs and the Grand Canal.

fēiwùzhì wénhuà yíchǎn

非物质文化遗产

Intangible Cultural Heritage

各族人民世代相传并视其为文化遗产组成部分的各种传统文化表现形式，以及与传统文化表现形式相关的实物和场所。包括传统口头文学以及作为其载体的语言，传统美术、书法、音乐、舞蹈、戏剧、曲艺和杂技，传统技艺、医药和历法，传统礼仪、节庆等民俗，传统体育和游艺等。截至 2022 年 4 月，北京有 144 项国家级代表性项目，303 项市级代表性项目。

Intangible cultural heritage (ICH) refers to traditional cultural expressions that have been handed down from generation to generation by people of all ethnic groups and are regarded as part of their cultural heritage, as well as objects and places related to traditional cultural expressions. It includes traditional oral literature and the language serving as its medium, traditional art, calligraphy, music, dance, drama, folk musical art forms and acrobatics, other traditional skills, medical and calendrical knowledge, traditional rituals, festivals and other folk customs, and traditional sports and amusements. As of April 2022, there are 144 national ICH items and 303 municipal ICH items in Beijing.

wénwù

文物

Cultural Relics

人类创造的或者与人类活动有关的，具有历史、艺术、科学价值的物质文化遗产。文物包括不可移动文物（古文化遗址、古墓葬、古建筑、石窟寺、石刻、壁画、近代现代重要史迹和代表性建筑等）和可移动文物（历史上各时代重要实物、艺术品、文献、手稿、图书资料、视听资料、代表性实物等）。截至 2022 年 4 月，北京有全国重点文物保护单位 135 项（189 处），市级文物保护单位 255 项（259 处）。

Cultural relics refer to heritage objects of historic, artistic and scientific value created by humans or related to human activities. They include immovable physical remains (such as archaeological sites and ruins, tombs, buildings of antiquity, cave temples, stone carvings, murals, historic sites from early modern and modern times and representative architecture) as well as movable items (such as important objects, artworks, documents, manuscripts, books, audiovisual materials, and representative items from all historical eras). As of April 2022, Beijing has 135 officially protected sites at the national level (at 189 locations) and 255 officially protected sites at the municipal level (at 259 locations).

gémìng shǐjì

革命史迹

Revolutionary Sites

自 1840 年以来，在中国人民实现国家独立、民族解放和民主自由以及建设社会主义现代化中国的历史实践过程中形成的重要遗产，既包括革命文物、各种建筑物和构筑物等物质遗产，也包括革命故事和歌曲等非物质遗产。截至 2022 年 4 月，北京已公布不可移动革命文物 158 处，可移动革命文物 2,111 件（套）。

Sites associated with revolutionary history constitute important heritage that was formed in the course of the Chinese people's practice in history since 1840 of pursuing national independence, liberation, democracy and freedom, and building a modern socialist country. They include both tangible heritage such as revolutionary heritage objects, buildings and structures, and intangible heritage such as revolutionary stories and songs. As of April 2022, Beijing has promulgated 158 immovable revolutionary heritage sites and 2,111 movable revolutionary heritage objects.

lìshǐ jiànzhù

历史建筑

Historic Buildings

能够反映历史风貌和地方特色，具有一定保护价值，尚未公布为文物保护单位且尚未登记为不可移动文物的建筑物、构筑物。原则上，建成时间应超过 50 年。截至 2022 年 4 月，北京市已公布历史建筑 1,056 栋（座），可分为合院式建筑、近现代公共建筑、工业遗产、居住小区和其他建筑五类。

Historic buildings are such structures that in principle are at least 50 years old, reflect historical features and local characteristics, are worth conservation, but have not been named the officially protected sites and have not been registered as immovable cultural heritage. As of April 2022, Beijing Municipality has announced 1,056 such historic buildings. They can be divided into five categories: courtyards, public buildings from early modern and modern times, industrial buildings, residential buildings and other buildings.

yōuxiù jìnxiàndài jiànzhù

优秀近现代建筑

Excellent Buildings of Early Modern and Modern Times

19 世纪中期至 20 世纪 70 年代期间建造的，现状遗存保存较为完整的，能够反映北京近现代城市发展历史，且具有较高历史、艺术和科学价值的建筑物（群）、构筑物（群）和历史遗迹。《北京历史文化名城保护条例》（2021 年）将优秀近现代建筑纳入历史建筑范畴。

Excellent buildings of early modern and modern times are those well preserved buildings and structures that date from the mid-19th century to the 1970s. Such buildings reflect Beijing's urban development in recent times and have substantial historic, artistic and scientific value. The "Regulations on the Conservation of the Historic City of Beijing" (2021) identify these buildings under the category of Historic Buildings.

shí dà jiànzhù

十大建筑

Ten Great Buildings

特指建成于1959年向国庆十周年献礼的北京十座公共建筑，是反映城市建设重要成果、体现时代精神的标志性建筑作品。“十大建筑”包括人民大会堂、中国革命博物馆和中国历史博物馆（今中国国家博物馆）、中国人民革命军事博物馆、民族文化宫、民族饭店、全国农业展览馆、钓鱼台国宾馆、华侨大厦、北京火车站、北京工人体育场，反映了新中国成立初期中国建筑艺术与技术的最高水平，是首都建筑史上的里程碑。

This term refers to Beijing's "ten great buildings" built in 1959 to commemorate the 10th anniversary of the People's Republic of China. They are remarkable architectural works that reflect the important achievements of urban construction and the spirit of the times. They are the Great Hall of the People, the Museum of the Chinese Revolution together with the Museum of Chinese History (now the National Museum of China), the Military Museum of the Chinese People's Revolution, the Cultural Palace of Nationalities, the Minzu Hotel, the National Agriculture Exhibition Center, the Diaoyutai State Guesthouse, the Overseas Chinese Hotel, the Beijing Railway Station, and the Beijing Workers' Stadium. Built to the highest level of Chinese architectural art and technology in the early years of the People's Republic of China, these buildings are milestones in the history of the capital's architecture.

èrshí shìjì jiànzhù yíchǎn

20 世纪建筑遗产

Twentieth-Century Architectural Heritage

1901—2000 年建成、对推动城市发展及社会进步具有重大作用的建筑，它们见证了人类社会变迁中最剧烈、最迅速的发展进程，与古代遗产具有同等重要的保护价值。“中国 20 世纪建筑遗产”评选由中国文物学会和中国建筑学会发起，截至 2022 年 4 月，已公布五批 497 处，其中北京有 94 处。

Twentieth-century architecture has played a major role in promoting urban development and social progress, evolved during the most dramatic and rapid development process in the change of human society, and is as worthy of conservation as ancient heritage. The selection of “China’s Twentieth-Century Architectural Heritage” was initiated by the China Cultural Relics Academy and the Architectural Society of China. As of April 2022, five groups of altogether 497 architectural heritage sites have been announced, including 94 in Beijing.

chuántǒng fēngmào jiànzhù

传统风貌建筑

Traditional-Style Buildings

历史文化街区内除文物和历史建筑之外具有一定建成历史，且能够反映历史风貌和地方特色的建筑。传统风貌建筑是历史文化街区的基底，其保护价值和措施不同于文物和历史建筑。

In addition to historic monuments and historic buildings, traditional-style buildings in historic conservation areas are those that were completed a certain number of years ago and reflect history and local characteristics. Such buildings constitute the basic setting of these areas and their conservation value and the conservation measures needed differ from those of historic monuments and historic buildings.

míngrén jiù(gù)jū

名人旧（故）居

Old (Former) Residences of Famous People

知名人士出生、长期生活工作，以及虽短暂居住但具有重要意义的居所。《北京历史文化名城保护条例》（2021 年）将名人旧居纳入历史建筑范畴。

Old or former residences are birthplaces of famous people or the homes where they lived or worked for a long time, or lived briefly but with great significance. The "Regulations on the Conservation of the Historic City of Beijing" (2021) include old residences of famous people in the category of Historic Buildings.

tèsè dìqū

特色地区

Areas of Special Quality

历史文化街区以外具有一定规模、代表一定时期城市规划建设探索实践、较为完整真实反映新中国成立初期北京城市风貌特色和社会生活的区域。截至 2022 年 4 月，北京有特色地区 13 片，包括百万庄小区、和平里小区、虎坊路小区、龙潭小区等。

Areas of special quality outside the historic conservation areas are those that are of a certain scale, represent exploratory practice in urban planning and construction of a certain period, and reflect the characteristics of Beijing's urban cityscape and social life in the early years of the People's Republic of China in a complete and truthful way. As of April 2022, there are 13 such areas in Beijing, including the Baiwanzhuang, Hepingli, Hufang Road and Longtan residential districts.

qítā chéngpiàn chuántǒng píngfángqū

其他成片传统平房区

Other Areas of Traditional Contiguous One-Story Houses

集中体现北京传统风貌、面积在 1 万平方米以上且尚未纳入历史文化街区的成片平房区，其与历史文化街区共同构成北京传统风貌的重要载体。

Other areas of traditional contiguous one-story houses are those that cover more than 1 hectare, are characteristic of Beijing's traditional townscape, and are not yet included in historic conservation areas. Together with the historic conservation areas, they constitute an important element of the traditional townscape.

dìxià wénwù máicángqū

地下文物埋藏区

Underground Archaeological Remains

根据史料、普查资料等划定的有可能集中埋藏文物的地区。截至2022年4月，北京市已公布地下文物埋藏区68处。

Underground archaeological remains are those that are designated as such in accordance with historical and general survey data and are likely to have concentrations of underground cultural objects. As of April 2022, Beijing has announced 68 such areas.

lìshǐ wénhuà míngzhèn

历史文化名镇

Historic Towns

保存文物特别丰富并且具有重大历史价值或者纪念意义，又能较完整地反映一定历史时期传统风貌和地方民族特色的乡镇。截至 2022 年 4 月，北京有中国历史文化名镇 1 处，即密云区古北口镇。

Historic towns are those that are particularly rich in historic monuments and also have great historic value or commemorative significance, as well as quite comprehensively reflecting the traditions of a certain historical period and local ethnic characteristics. As of April 2022, Beijing has one historic town, namely Gubeikou Town in Miyun District.

lìshǐ wénhuà míngcūn

历史文化名村

Historic Villages

保存文物特别丰富并且具有重大历史价值或者纪念意义，又能较完整地反映一定历史时期传统风貌和地方民族特色的村庄。截至 2022 年 4 月，北京有中国历史文化名村 5 处，即门头沟区的斋堂镇爨底下村、斋堂镇灵水村、龙泉镇琉璃渠村，以及顺义区龙湾屯镇焦庄户村、房山区南窖乡水峪村。

Historic villages are those that are particularly rich in historic monuments and also have significant historic value or commemorative significance, as well as quite comprehensively reflecting the traditions of a certain historical period and local ethnic characteristics. As of April 2022, Beijing has five national level historic villages, namely Cuandixia Village and Lingshui Village in Zhaitang Town, and Liuliqu Village in Longquan Town, all in Mentougou District, as well as Jiaozhuanghu Village in Longwantun Town, Shunyi District, and Shuiyu Village in Nanjiao Town, Fangshan District.

chuántǒng cūnluò

传统村落

Traditional Villages

建成超过百年、拥有物质形态和非物质形态文化遗产，且具有较高的历史、文化、科学、艺术、社会、经济价值的村落。截至 2022 年 4 月，北京有市级传统村落 44 处，其中 22 处被列入中国传统村落名录。

Traditional villages are those that were built up more than a century ago, have tangible and intangible cultural heritage, and high historic, cultural, scientific, artistic, social and economic values. As of April 2022, Beijing has registered 44 places as traditional villages, 22 of which are included in the List of Traditional Villages in China.

gōngyè yíchǎn

工业遗产

Industrial Heritage

在北京工业发展进程中形成的具有较高的历史、文化、科学、艺术、社会、经济价值的遗产，既包括建（构）筑物、设备和产品等物质遗产，也包括工艺流程和管理体系等非物质遗产。《北京历史文化名城保护条例》（2021 年）将工业遗产纳入历史建筑范畴。

Industrial heritage is heritage with high historic, cultural, scientific, artistic, social and economic values created in the process of industrial development in Beijing, including both tangible heritage such as buildings, structures, equipment and products, and intangible heritage such as processes and management systems. The "Regulations on the Conservation of the Historic City of Beijing" (2021) include industrial heritage in the category of Historic Buildings.

shuǐwénhuà yíchǎn

水文化遗产

Aquatic Cultural Heritage

与城市沿革密切相关且具有历史文化价值的井、桥、闸、坝河和码头等水工建筑物、构筑物和遗址。北京水文化遗产资源丰富，如大运河沿线的张家湾码头、南新仓、万宁桥（澄清上闸）、白浮泉等，永定河沿线的卢沟桥、庞村古堤等。

Aquatic cultural heritage refers to buildings, structures and sites related to waterworks such as wells, bridges, sluice gates, dams, rivers and docks that are closely connected with urban development and have historic and cultural values. Beijing is rich in such heritage, such as the Zhangjiawan Pier, Nanxincang Granary, Wanning Bridge (Chengqing Upper Sluice Gate) and the Baifuquan Spring along the Grand Canal, and the Lugou Bridge and Pangcun Ancient Dike along the Yongding River.

lìshǐ hé hú shuǐxì

历史河湖水系

Historic Rivers and Lakes

与城市沿革密切相关且具有历史文化价值的河道或水域。北京历史河湖水系资源丰富，老城内有“六海八水”，老城外有永定河、潮白河、温榆河等。

Historic rivers and lakes are those that are closely related to a city’s history and also have cultural values. Beijing is rich in such water resources. There are the “six lakes and eight rivers” in the Old City as well as the Yongding, Chaobai and Wenyu rivers outside the Old City.

shānshuǐ géjú

山水格局

The Mountain-and-River-Based Pattern

在城市选址及规划建设过程中，充分考虑城市与区域自然山水环境的融合，以特定的轴线或视线形成了具有意境的“山—水—城”格局，反映了中国天人合一的哲学理念。北京的“大山大水”包括燕山山脉、太行山山脉、永定河、拒马河、潮白河、沟河等。“左环沧海，右拥太行，北枕居庸，南襟河济”（《大明一统志》）形象地勾勒出北京的山水格局。

The mountain-and-river-based pattern comes into being by integrating the city with the natural environment of mountains and rivers in site selection and planning process, with a specific axis or line of sight featuring an artistic mountain-river-city layout and reflecting the Chinese philosophic concept of harmony between humans and nature. Beijing's mountains and rivers include the Yanshan and Taihang ranges and the Yongding, Juma, Chaobai, and Ju rivers. This mountain-river pattern is vividly outlined (as seen from the north) in the *Comprehensive Gazetteer of the Great Ming*: “To the left girdled by the blue sea, to the right held by the Taihang range, in the north resting against Juyong Pass, in the south embraced by the Yellow and Ji rivers.”

fēngjǐng míngshèng

风景名胜

Scenic Areas

以名胜古迹为主体，自然景观与人文景观和谐相融，兼具有观赏、文化或者科学价值并可供人们游览或进行科学、文化活动的区域。如北京的八达岭—十三陵风景名胜区。

Scenic areas are notable for their beautiful natural environment and historical remains. Harmoniously integrating the natural and built environment, they have aesthetic, cultural or scientific values and can be visited for sightseeing or for scientific and cultural activities. In Beijing they include the Great Wall at Badaling and the Ming Tombs.

chéngzhǐ yícún

城址遗存

City-Site Remains

历史上形成的，具有较高价值的古城遗迹、生产生活及活动痕迹等的遗产区域。北京现存多个历史时期的城址遗存，如西周燕都琉璃河遗址，辽南京、金中都、元大都、明清北京城城址遗存等。

City-site remains are heritage areas of high value with ruins and traces of daily life, economic activity, etc. At present Beijing has such sites from multiple historical periods, such as the site of the capital of Yan in the Western Zhou period at Liulihe, the remains of the Liao Dynasty's Nanjing (Southern Capital), the Jin Dynasty's Zhongdu (Central Capital), the Yuan Dynasty's Dadu (Great Capital), and the Ming and Qing dynasties' city of Beijing.

guójiā wénhuà gōngyuán

国家文化公园

National Cultural Parks

将公众高度认同、能够代表国家形象和中华民族独特精神标识、独一无二的文物和文化资源进行系统整合，并实施公园化管理运营的综合性文化空间。国家文化公园具有国家象征意义，有助于公众了解、体验中国历史和中华文化，促进区域经济、社会和生态建设协调发展。目前，我国已有长城、大运河、长征、黄河四大国家文化公园。其中，长城国家文化公园和大运河国家文化公园涉及北京。

National cultural parks are comprehensive cultural spaces that systematically integrate unique historic monuments and cultural resources that are well recognized by the public, represent the image of China and the unique spiritual identity of the Chinese nation, and are managed and operated as parks. National cultural parks have multiple functions such as being of holistic national significance, providing the public with the opportunity to understand and experience Chinese history and culture, and promoting the coordinated economic, social and ecological advances in different regions. Currently, China has four such parks: the Great Wall, the Beijing-Hangzhou Grand Canal, and the Long March and Yellow River national cultural parks. The first two of these parks are partly in Beijing.

yízhǐ gōngyuán

遗址公园

Parks of Ruins

以重要考古遗址及其背景环境为主体，具有科研、教育、游憩等功能，在考古遗址保护和展示方面具有示范意义的特定公共空间。北京第一处遗址公园是1985年建成的团河行宫遗址公园。此后又陆续建成圆明园遗址公园、元大都城垣遗址公园、皇城根遗址公园及明城墙遗址公园等。圆明园遗址公园和周口店遗址公园被列入首批国家考古遗址公园。

Dominated by important archaeological sites and their surroundings, parks of ruins are specific public spaces with functions of scientific research, education, recreation, etc. They play an important exemplary role in the conservation and display of archaeological sites. The first such park in Beijing is the Tuanhe Palace Ruins Park, established in 1985. Since then, Yuanmingyuan Ruins Park, Yuan Dadu City Wall Ruins Park, Huangchenggen Ruins Park, and the Ming City Wall Ruins Park have been successively established. Yuanmingyuan Ruins Park and Zhoukoudian Ruins Park are two of China's first batch of national-level archaeological ruins parks.

chuántǒng hútòng

传统胡同

Traditional *Hutongs*

清代及以前形成的、基本保留原有宽度及走向、历史风貌尚存且具有一定保护价值的胡同。胡同与四合院共同构成北京典型的传统空间形态。北京现存各时期形成的传统胡同 1,000 余条，其中，北京有文字记载最早的是砖塔胡同和史家胡同，始自元代。

Traditional *hutongs* are those dating from the Qing Dynasty or earlier that basically retain their original width, direction and historical features, and are worth preserving. Such *hutongs* together with *siheyuan* compounds constitute the typical traditional spatial formations of Beijing. There are more than 1,000 such traditional *hutongs* in Beijing from more than one historical period, the earliest recorded ones being Zhuanta and Shijia *hutongs*, dating back to the Yuan Dynasty.

lìshǐ jiēxiàng

历史街巷

Historic Streets and Alleys

20 世纪初期至 20 世纪 70 年代期间经整体规划形成的街巷，如北京的香厂路、万明路、右安门大街、真武庙头条、南礼士路等。

Historic streets and alleys are those created under integrated planning from the early 20th century to the 1970s, such as Xiangchang Road, Wanming Road, You’anmen Street, Zhenwumiao Toutiao Alley and Nanlishi Road in Beijing.

chuántǒng dìmíng

传统地名

Traditional Place Names

各历史时期流传下来或者曾使用过的地名，以街巷和胡同名称为主，也包括古城、古镇、古村等名称。截至 2022 年 4 月，北京已公布第一批传统地名（“街巷胡同类”）598 处。

Passed down from or used in various historical periods, traditional place names are mainly street, alley and *hutong* names but also include names of old cities, towns and villages. As of April 2022, Beijing has announced the first list of 598 traditional names for streets, alleys and *hutongs*.

lìshǐ míngyuán

历史名园

Historic Gardens

是指具有突出的历史、文化、生态、科学价值，能体现特定历史时期造园技艺，对城市变迁或文化艺术发展产生过影响的园林场景。截至 2022 年 4 月，北京已公布历史名园 25 座，如北海公园、景山公园、恭王府花园等。

Historic gardens are those of high historic, cultural, ecological and scientific values that reflect gardening skills of specific historical periods, and have exerted a certain influence on urban evolution or cultural and artistic development. As of April 2022, Beijing has designated 25 historic gardens, such as Beihai Park, Jingshan Park and Prince Kung's Palace Garden.

gǔshù míngmù

古树名木

Ancient and Valuable Trees

“古树”指树龄在 100 年以上的树木，其中树龄在 300 年以上的为一级古树，其余为二级古树。“名木”指珍贵、稀有的树木和具有历史价值、科学价值、纪念意义的树木。据普查，目前北京各级各类古树名木共 41,865 株，其中位于密云区新城子镇的侧柏“九搂十八杈”是已知最古老的树木，树龄约 3,500 年。

“Ancient trees” are those that are 100 years old or more, of which those aged 300 years or more are first-class ancient trees and the rest second-class. “Valuable trees” refer to precious and rare trees and those of historic or scientific value, or commemorative significance. A recent tree census shows that there are currently 41,865 trees of both classes in Beijing, among which the Eighteen-Bough Ancient Cypress, located in Xinchengzi Town, Miyun District, and estimated to be about 3,500 years old, is the oldest known tree in Beijing.

lǎozìhao

老字号

Time-Honored Brands

历史悠久、拥有世代传承的产品、技艺或服务，具有鲜明的中华民族传统文化背景和深厚的文化底蕴，取得社会广泛认同，形成良好信誉的品牌。北京众多老字号享誉国内外，涉及餐饮、零售、食品、医药等行业，如全聚德、盛锡福、吴裕泰、同仁堂等。

Time-honored brands with a long history are those for products, skills or services that have been passed down from generation to generation. They should have a distinctive traditional Chinese cultural background and represent a deep-rooted heritage, be widely recognized by the public and have a fine reputation. Many such brands in Beijing are well known at home and abroad. Found in catering, retail, foodstuffs, medicine and other industries, these brands include Quanjude, Shengxifu, Wuyutai and Tongrentang.

chuántǒng gōngyì

传统工艺

Traditional Crafts and Techniques

世代相传、具有百年以上历史以及完整工艺流程，采用天然材料制作且具有鲜明民族风格和地方特色的工艺品种及其技艺。北京流传至今的传统工艺是宝贵的非物质文化遗产，如官式古建筑营造技艺、剧装戏具制作技艺、京绣、琉璃烧制技艺、景泰蓝制作技艺、北京绢花等。

Traditional crafts and techniques are those that have been handed down for generations over a period of more than 100 years and represent complete processes. They should use natural materials and have uniquely ethnic and local characteristics. Such crafts passed down to this day in Beijing are a valuable intangible cultural heritage. They include the techniques employed in building government offices, making stage costumes and props, Beijing embroidery, glaze-firing techniques, cloisonné production techniques, and Beijing silk flowers.

chuántǒng cáiliào

传统材料

Traditional Materials

用于传统建筑建造的材料。如建造故宫所使用的金砖、木材、琉璃瓦等，建造传统四合院所使用的青砖、灰瓦、小青石、石灰等。

Used in the construction of old-style buildings, traditional materials include the solid, square bricks (*jinzhuan* 金砖, specially made for paving floors of the imperial palaces, and when they are hit clanging like that of metal being hit; "metal" in Chinese called "金属," hence the term "金砖"), wood, and glazed tiles used in the construction of the Forbidden City, and the gray bricks and tiles, small blueish-gray stones and lime used in the construction of *siheyuan* compounds.

chuántǒng guīzhì

传统规制

Traditional Regulations

传统建筑在规模形制、用材用料等级方面的规定。中国古代对建筑的屋顶式样、用材等级、面阔间数、台基层数、装饰色彩等诸多方面都有具体规定。规制代表着建筑的等级和规格，是中国传统礼制的物质表现。如故宫太和殿的重檐庑殿顶、面阔十一间、三层汉白玉石雕台基等特征，都是最高等级皇家建筑的体现。

China had specific regulations on many aspects of buildings such as the style of the roof, the grade (quality) of materials, the width and number of rooms, the number of steps up to building floor, and the colors of decorations. Specifications regarding the design and standard of buildings are an expression of China's traditional ritual system. For example, the features of the Forbidden City's Hall of Supreme Harmony, such as hipped roof with double eaves, a façade with a width of 11 bays (*jian*), and a three-level foundation of white marble, are the embodiment of the highest grade of imperial architecture.

lìshǐ chéngqū

历史城区

Historic Urban Areas

城镇中能体现其历史发展过程或某一历史时期风貌的地区，且具有历史范围清楚、格局和风貌保存较为完整的特点。北京有多处不同时期形成的历史城区，如北京老城、通州古城、路县故城、宛平城、永宁古城等。

Historic urban areas are those that reflect their process of development over time or character in a specific historical period. Such areas are characterized by a clear historical boundary and a well-preserved layout and appearance. Beijing has multiple historic urban areas formed in different periods, such as the Old City of Beijing, Tongzhou Old Town, Luxian Old Town, Wanping City, and Yongning Old Town.

lìshǐ géjú

历史格局

Historic Patterns

由历史城垣轮廓、河湖水系、街巷肌理、重要节点等组成的具有保护价值的空间组合形式，是北京传统营城理念的外在表现。两轴、四重城廓、六海八水、九坛八庙、棋盘路网是老城空间格局的重要特征，是奠定老城空间地位的重要载体，加强格局保护是老城整体保护最重要的任务之一。

Historic patterns refer to spatial combinations worth preserving, such as city walls, rivers and lakes, streets and alleys, and important intersections. These are the external expression of Beijing's traditional city-construction concepts. The two axes, the quadruple-walled city framework, the six lakes and eight rivers, the nine altars and eight temples, and the orthogonal grid street layout are the important features of the Old City's spatial pattern and an important means of establishing its spatial status. Strengthening the protection for this pattern is one of the most important tasks in the overall conservation of the city.

lìshǐ huánjìng yàosù

历史环境要素

Historic Environment Elements

反映历史风貌的古井、围墙、石阶、铺地、水系、驳岸、古树名木等，是历史城区、历史文化街区等保护对象所在历史环境的有机组成，应与保护对象实施同步保护。

Historic environment elements include ancient wells, protective walls, stone steps, paved areas, water systems, revetments, ancient and valuable trees, etc. which reflect historic features. These are integral parts of the environment in historic urban areas and historic conservation areas. They are to be protected along with the other objects under conservation.

chuántǒng gōngjiàng

传统工匠

Traditional Craftsmen

具有高度职业精神、从事传统建筑修缮并掌握高超技艺的从业人员，是传统文化的传承人。明清时期古建筑有八大作，包括土作、石作、搭材作、木作、瓦作、油作、彩画作、裱糊作。

Traditional craftsmen are highly professional people who are engaged in the preservation of traditional-style buildings and are masters with superb skills who perpetuate traditional culture. There are eight main crafts of ancient Ming-Qing dynasty architecture, including work with earth, stone, scaffolding, wood, tiles, oil painting (for protecting wood), wood decorating, and wallpaper.

第三篇
保护要求与保护措施
Part III
Requirements and Measures for Conservation

lǎochéng bù néng zài chāi

老城不能再拆

No More Demolition in the Old City

不能再通过大拆大建、推倒重来、拆旧建新等方式实施老城更新，要通过腾退、恢复性修建等，做到应保尽保。“老城不能再拆”是对老城整体保护提出的战略要求。

There must be no more demolition in the Old City through large-scale renewal by pulling down old and putting up new structures, but instead there should be such measures as evacuation and restoration to preserve as much as possible. “No more demolition” is a strategic requirement for the comprehensive conservation of the Old City.

bù qiú suǒ yǒu, dàn qiú suǒ bǎo, xiàng shèhuì kāifàng

不求所有、但求所保，向社会开放

Protect Heritage Regardless of Ownership and Give Access to the Public

不以产权归属作为对文物等保护对象履行保护义务的前提，提倡产权人、使用人和有关方面共同做好保护工作，并在此基础上向社会开放。

Ownership must not be a prerequisite for the conservation of historic monuments, while property owners, users and other stakeholders should be encouraged to protect such heritage. On this basis it shall be open to the public.

yīng bǎo jìn bǎo

应保尽保

Protect All that is Worth Protecting

通过对历史文化价值、建筑艺术价值、科学技术价值等的深入挖掘，确保具有保护价值的遗存都能得到保护。“应保尽保”体现了一种积极的保护态度，对于做好老城整体保护具有重要意义。

Everything must be done to protect old remains that are worth preserving through in-depth identification of their historical and cultural values, architectural and artistic values, and scientific and technological values. The principle of "protecting all that is worth protecting" expresses a positive attitude to conservation and is of great significance for the comprehensive conservation of the Old City.

zhěngtǐ bǎohù

整体保护

Comprehensive Conservation

以系统思维对保护对象、依存环境及活态遗产和社区结构等实施全要素保护和全过程保护，实现在保护中发展、在发展中保护。在北京老城整体保护中，除保护文物、历史建筑、历史文化街区外，还要加强对历史格局、景观视廊、城市风貌等的保护。

Comprehensive conservation means adopting a systemic approach to the all-key-factors and whole-process conservation of the elements to be protected, the environment they depend on for survival, the living heritage, and the community fabrics, so as to realize development in the process of conservation and conservation in the process of development. In the comprehensive conservation of Beijing's Old City, in addition to the conservation of historic monuments, historic buildings and conservation areas, the conservation of historic patterns, visual corridors and city features must also be strengthened.

zhēnshíxìng

真实性

Authenticity

文化遗产本身的形态、工艺、设计及其环境和它所反映的历史、文化、社会等相关信息的真实性。对文化遗产的保护就是保护这些信息及其来源的真实性，与其相关的文化传统的延续也是对真实性的保护。

The authenticity of the cultural heritage means its form, craftsmanship, design and setting and the historical, cultural, social and other related information that it reflects are credible and truthful. Safeguarding cultural heritage means protecting the above-mentioned information and its sources, and ensuring the continuity of the cultural traditions associated with it.

wánzhěngxìng

完整性

Integrity

文化遗产核心价值、价值载体及其环境等要素应得到完整保护。文化遗产在历史演化过程中形成的包括各个时代特征、具有价值的物质遗存都应得到尊重。

Elements of cultural heritage such as core values, value carriers and their setting should receive comprehensive protection. The valuable material remains of cultural heritage, including the characteristic features of each era, formed during historical evolution, should be respected.

yánxùxìng

延续性

Continuity

对保护对象适合当代社会生活功能的延续。对于具有活态特征的保护对象，应延续原有功能，并保护其具有文化价值的传统生产、生活方式。

Continuity means preserving the functions of the conservation objects in ways suited to modern social life. For such living heritage, its original functions shall be continued and traditional ways of production and life with cultural value shall be protected.

xiǎo guīmó, jiànjìnshì

小规模、渐进式

Progressive Adaptation

以解决使用者实际问题为目的的更新方式，也称“微更新”。较之以往以成片推倒重建为基本特征、统一设计、统一建设的大规模改造方式，微更新有利于最大限度保护街区的历史人文环境和风貌特色，更易于满足居民的多样化需求。

Also known as “micro-renewal,” progressive adaptation aims to solve the practical problems of residents. Unlike the previous practice of large-scale urban renewal characterized by demolishing and rebuilding the whole area with unified design, micro-renewal is conducive to maximizing the conservation of the historical and cultural environment and the features of neighborhoods, and makes it easier to meet the diversified needs of residents.

shūjiě zhěngzhì cù tíshēng

疏解整治促提升

Rehabilitation for Improvement

立足首都城市战略定位，疏解非首都功能，有效治理“大城市病”，大力改善人居环境，全面提升城市品质，不断增强发展活力。重点任务包括一般性产业疏解提质，公共服务功能疏解提升，违法建设治理与腾退土地利用，桥下空间、施工围挡及街区环境秩序治理，重点区域环境提升等。

This policy refers to relieving Beijing of such activities that are nonessential to a capital city, dealing effectively with “big city maladies,” bettering the living environment, vigorously improving the city in every aspect, and continuing to invigorate its development. Key tasks include relocating the general industry, facilitating and improving public service functions, addressing illegal construction and misuse of land, and improving space under bridges, construction sites, the conditions and orderliness in urban neighborhoods, and the environment in key areas.

xiǎoxíng zōnghé guǎnláng

小型综合管廊

Small Utility Tunnels

适用于历史文化街区等空间资源紧张路段，对电力电缆、通信线缆、给水配水管道、再生水配水管道等实施统一规划、设计、建设和管理，内部空间可不考虑人员通行的小规模、集约化城市管道综合走廊。

In historic conservation areas, small-scale integrated utility tunnels are used in areas with limited space. They make possible unified planning, design, construction and management of power cables, communication cables, water supply pipes, and pipes for distribution of reclaimed water. Passage of personnel is not a requirement for such tunnels.

huóhuà lìyòng

活化利用

Activation and Utilization

统筹做好保护、传承、利用工作，推动文物和历史建筑等从静态保护向积极保护转变，实现历史文化遗产的创造性转化和创新性发展。

This term is used to call for coordinated efforts to protect, utilize and pass on to future generations historic monuments and historic buildings in order to promote the transformation from static to active conservation, and to realize the creative transformation and innovative development of historic and cultural heritage.

wénhuà tànfǎng lù

文化探访路

Culture Roads

以街巷为依托，将历史文化街区、文物古迹、现代文化设施和城市景点用舒适的慢行环境串联起来的文化线路，沿途提供方便的服务设施，满足人们观光游览、历史寻踪、文化探访和城市体验的需求。

Culture roads are made up of streets and alleys that link historic conservation areas, historic monuments and sites, modern cultural facilities and urban attractions in an environment suitable for comfortable strolling. Convenient service facilities are provided along the way for sightseeing, tracing history, and cultural experience and urban exploration.

bǎohù guīhuà

保护规划

Conservation Planning

以保护历史文化遗产、协调保护与建设发展为目的，以确定保护的原则、内容和重点，确定保护对象、划定保护区划、提出保护管理措施为主要内容的规划。保护规划是国土空间规划中的专项规划。历史文化名城、历史文化街区、历史文化名镇名村、文物保护单位和历史河湖水系等各类保护对象均可编制相应的保护规划。

Conservation plans are aimed at preserving historic and cultural heritage and coordinating conservation, construction and development. They mainly focus on determining the principles, contents and priorities of conservation, identifying objects to be preserved, delineating conservation zones and proposing conservation and management measures. Conservation plans are a special part of the territorial spatial planning. Appropriate conservation plans may be worked out for historic cities, historic conservation areas, historic towns and villages, the officially protected monuments and sites, historic rivers and lakes, and other types of conservation objects.

bǎohù qūhuà

保护区划

Conservation Zoning

根据保护对象的价值和保护该价值载体而划定的保护管理区域，由保护范围和建设控制地带构成，通过制定严格具体的保护管理要求来防范各类建设活动对保护对象可能造成的负面影响。

Conservation and management zones are delineated in accordance with the value of the objects to be preserved and the need to protect the value carriers. They are composed of conservation areas and construction-control zones. Strict and specific conservation and management requirements are set to prevent the possible negative effect of various construction activities on the conservation objects.

(wénwù hé lìshǐ jiànzhù) bǎohù běntǐ

（文物和历史建筑）保护本体

Fabric of Historic Monuments and Historic Buildings

文物建筑和历史建筑中承载核心价值的物质构成，是开展保护、管理、展示工作的重点。保护本体、历史格局、古树名木、非物质遗存等共同构成了文物建筑和历史建筑的价值。

Fabric is the material remains that constitute the core value of historic monuments and historic buildings. They have priority of conservation, management, and display work. The historic fabric, historic patterns, ancient and valuable trees, intangible cultural heritage and so on constitute the value of these buildings.

wénwù bǎohù dānwèi bǎohù fànwéi

文物保护单位保护范围

The Conservation Area of the Officially Protected Sites

对文物本体及周围一定范围实施重点保护的区域，以满足对文物古迹格局、安全、环境和景观的保护要求。文物保护范围应当根据文物保护单位的类别、规模、内容以及周围环境的历史和现实情况合理划定，以确保对文物本体及其环境的有效保护。

The conservation area of the officially protected sites is the area specified as the focus of protection for the sites themselves as well as a certain range around them in order to meet the conservation requirements for the configurations, security and setting of the heritage sites. The extent of the protected sites shall be reasonably delimited in accordance with their categories, scale and attributes as well as the historical and current conditions of their surrounding environment, so as to ensure effective conservation of the sites themselves and their environment.

wénwù bǎohù dānwèi jiànshè kòngzhì dìdài

文物保护单位建设控制地带

Construction-Control Zones for the Officially Protected Sites

在文物保护单位的保护范围外，为保护文物保护单位的安全、环境、历史风貌，对建设项目加以限制的区域。北京文物保护单位周围的建设控制地带分为五类，其中一类至四类地带管控强度依次递减；五类地带为特殊控制地带，需针对文物保护单位的特殊价值或特殊情况明确相应管控要求。

Construction-control zones exist outside the scope of conservation in areas where building projects are restricted to protect the safety, environment and historical features of the officially protected sites. Such construction-control zones in Beijing are divided into five categories, among which the intensity of control decreases successively from Grade I to Grade IV, and the zone with Grade V is a special zone for which control requirements shall be specified in accordance with the special value or special circumstances of the officially protected sites.

wénwù jiànzhù yùfángxìng bǎohù

文物建筑预防性保护

Preventive Conservation of Historic Monuments

预先对文物建筑及所处环境采取监测、检查、保护、加固和其他管理措施，避免潜在危害因素可能造成的损害。

Preventive conservation of historic monuments means taking monitoring, inspection, conservation, stabilization and other management measures in advance for historic monuments and the environment in which they are located to avoid damage caused by potential hazards.

wénwù jiànzhù xiūshàn

文物建筑修缮

Historic Monument Repair

以排除险情、改善文物本体状况、保护历史文化价值为目的的修缮活动。在修缮过程中，应尽量保留原有构件，修补、修复的部分应当可识别，并与原有风格一致。

Historic monument repair is activity aimed at eliminating hazards, improving the condition of the historic fabric and preserving historic and cultural values. In the repair process, the original components must be retained as far as possible, and the repaired and restored parts are to be recognizable and consistent with the original style.

làojià dàxiū

落架大修

Restoring the Wooden Structure Through Complete Disassembly

一种对木结构文物建筑干预强度最大的传统维修方法。在文物建筑主要结构严重变形或者主要构件受到严重损伤，非解体不能恢复安全稳定时，可以对其进行局部或全部解体。解体修复后应排除所有安全隐患，并确保在较长时间内不再修缮。

Restoration through complete disassembly is a traditional repair approach for historic wooden monuments needing the most intervention intensity. When the main structure of such a building is seriously deformed or its main components are badly damaged so that safety and stability cannot be restored without taking it apart, it may be partially or fully disassembled. Such action should remove all safety hazards and ensure that no further repair work is necessary for a long period of time.

wénwù téngtuì

文物腾退

Vacating Historic Monuments

一种制止和纠正文物建筑不合理使用的措施。由实施主体与使用方签订腾退协议，并收回文物建筑的行为。腾退后的文物建筑经修缮保护后活化利用，向社会开放。

Vacating historic monuments is a measure to stop and correct the unreasonable use of such buildings. It is implemented by signing an evacuation agreement with the user or users to recover the buildings. After vacated historic monuments are repaired, they shall be brought back into use and opened to the public.

kǎogǔ kāntàn

考古勘探

Archaeological Exploration

通过钻探、物探等手段来了解、确认和研究文化遗存，并为考古发掘和文化遗产保护提供基础材料与依据。最常用的考古勘探方法是使用探铲（又称“洛阳铲”）定位夯土、基址等遗存。

The aim of archaeological exploration is to understand, identify and study cultural heritage through drilling, physical exploration and other means, and to provide basic materials and support for archaeological excavations and cultural heritage conservation. The most common method of such exploration is the use of the shovel (also known as the "Luoyang shovel") to locate the remains of rammed earth and building foundations.

kǎogǔ fājué

考古发掘

Archaeological Excavation

为了科学研究，对地下埋藏文物的地方，如古文化遗址、古墓葬等，进行调查、勘探、发掘活动。一切考古发掘工作，必须依法履行报批手续。从事考古发掘的单位和有关发掘计划，均需国务院文物行政部门批准。

Archaeological excavation is done for scientific research and involves the investigation, exploration and excavation of the places with antiquities and monuments buried underground (e.g. ancient cultural sites and tombs). All excavation work requires approval through legal procedure. Archaeological institutions and relevant excavation plans must be approved by the cultural heritage administration under the State Council.

lìshǐ jiànzhù rìcháng bǎoyǎng

历史建筑日常保养

Routine Maintenance of Historic Buildings

以消除历史建筑安全隐患或者维护环境整洁为目的的日常和周期性措施。主要包括修补破损的瓦面，原样更换破损门、窗、台阶，防渗、防潮、防蛀、防漏，清除影响历史建筑安全的杂草植物等，在此过程中应注意重点保护好体现其核心价值的外观、结构和构件等。

The everyday maintenance of historic buildings is a daily or periodic measure for eliminating potential safety hazards and keeping the environment clean and tidy. It mainly includes repairing broken tiles, replacing damaged doors, windows, steps and stairs with new ones of the same type, preventing insects, seepage, moisture and leaks, and removing weeds and plants affecting building safety. During such works, attention shall be paid to the conservation of the appearance, structure and components of the buildings that reflect their core value.

lìshǐ jiànzhù xiūshàn

历史建筑修缮

Historic Building Rehabilitation

在保护历史建筑核心价值的前提下，为适应当代功能要求，对造成损坏的根源和安全隐患行为进行排除的措施。需依据建筑立面、结构体系、平面布局、特色构件和历史环境要素等有价值部位的保存情况，确定具体的修缮要求。国家鼓励活化利用历史建筑，在保持原有外观风貌、典型构件的基础上，通过加建、改建、添加或改造设施等方式适应现代生产生活需要。

This is done on the precondition of protecting the core value and in accordance with present-day functional requirements. The aim is to remove the root causes of damage and activities that are hazardous to safety. Specific repair requirements must be determined based on the conservation of valuable parts such as the façade, structural configuration, floor plans, distinguishing components and historic environment elements. The government encourages revitalization and utilization of historic buildings on the basis of maintaining the original appearance and typical components. This should be done by expanding, altering, adding or transforming facilities to adapt the buildings to the needs of modern life and production.

lìshǐ jiànzhù qiānyí

历史建筑迁移

Relocation of Historic Buildings

因重大公共利益需要，需使用历史建筑所在位置，不得已对历史建筑实施整体迁移或解体后迁移的行为。实施迁移前，应做好相关信息记录，并妥善保存解体后的建筑构件。

When overwhelming public interest requires the use of the locations of historic buildings, it becomes necessary to move them away in their entirety or to dismantle them for relocation. Before such removal, relevant information should be carefully recorded and disassembled building components should be well kept.

huīfùxìng xiūjiàn

恢复性修建

Restorative Construction

当历史格局和传统风貌已发生不可逆改变或无法通过修缮、改善等方式继续维持传统风貌时，以历史格局、传统形态、空间尺度研究为依据，辨析有保护价值的要素后开展的有条件、有限度的恢复建设行为。

Restorative construction is a kind of reconstruction with restraint under certain circumstances, which can only be carried out on sites where the historic pattern and traditional townscape have been irreversibly changed or where it is impossible to keep them through preservation and improvement. There is a need to study the earlier layout, style and spatial scale of such sites while identifying and conserving valuable elements.

shēnqǐngshì tuìzū

申请式退租

Application for Termination of Tenancy

老城历史文化街区指定片区内平房直管公房居民自愿向实施主体提出退租申请、签订申请式退租协议，经营管理单位根据协议与承租人解除租赁合同，收回直管公房使用权，实施主体给付承租人货币补偿。自愿退租的直管公房承租人领取货币补偿后，可申请区政府提供的共有产权房源和公租房房源。

Tenants of government-subsidized and public-owned one-story houses in specific sections of the historic conservation areas of the Old City of Beijing may voluntarily apply to the authorized management entity for termination of their tenancies and sign termination agreements. The management entity may agree to terminate the lease, pay the tenant compensation, and recover the right to use the relevant public housing. Besides receiving compensation, the tenant may apply for joint-ownership housing or public rental housing provided by the local government.

shēnqǐngshì huànzū

申请式换租

Application for Rent Exchange

在实施历史文化街区平房直管公房片区内，换租人自愿提出申请，与换租主体协商，可通过换租主体提供租赁房源置换换租人房屋使用权或换租主体市场化租赁换租人房屋使用权两种方式，改善居住条件。在“申请式退租”的基础上增加“申请式换租”模式作为补充，旨在为居民改善生活条件提供更多选择。

When tenants of government-subsidized and public-owned one-story houses in historic conservation areas apply for termination of tenancy, they also have the option of applying to improve their living conditions by asking the authorized management entity to either exchange their housing unit for another one or pay them rent based on market value if they agree to move out of the housing unit. The possibility of applying for change of tenancy as a supplementary measure when terminating a tenancy is intended to provide more choices for residents to improve their living conditions.

shēnqǐngshì gǎishàn

申请式改善

Application for Improvement of Living Conditions

在实施历史文化街区申请式退租、申请式换租片区内，选择继续留住平房区的居民可自愿选择实施主体提供的房屋改善菜单并承担部分费用的模式。

Tenants who choose to stay in historic conservation areas where the policies of application for termination of tenancy and application for rent exchange are applicable may choose optional services provided by the authorized management entity while shouldering part of the costs.

gòngshēngyuàn

共生院

Shared Courtyard

体现新老建筑共生、新老居民共生和文化共生的老城院落更新模式。老城平房院落内完成局部腾退并修缮后的建筑，一方面可补充基本生活配套设施，为留下来的居民改善居住条件，另一方面可引入新居民或者文创产业等，助力老城复兴。

The shared courtyard is intended to achieve a harmonious relationship between old and new buildings, long-time residents and newcomers, traditional and modern culture. The one-story courtyard houses that have been vacated and renovated completely or in part may either be furnished with basic facilities to improve the living conditions for residents who have stayed there, or else have new residents or cultural and creative enterprises introduced to help revitalize Beijing's Old City.

zérèn guīhuàshī zhìdù

责任规划师制度

The Mechanism of Designated Area Planners

为保持一定区域内城市规划的系统性、建筑风貌的协调性、文脉的可传承，聘任提供长期咨询服务工作的规划师与建筑师的工作机制。截至2021年年底，北京市333个街乡地区已实现责任规划师全覆盖，形成了约300个团队、上千人的责任规划师队伍，全面参与到研究、设计、审查、建设指导、后评估全过程，是规划落实“最后一公里”的重要力量。

This is a working mechanism of governments at all levels in Beijing to appoint planners and architects who provide long-term consulting services in order to maintain the systematic character of urban planning, harmonization of architectural styles, and the consistency of the cultural heritage in a certain area. By the end of 2021, 333 townships in Beijing had been fully covered by the mechanism of designated area planners. There are roughly 300 teams with altogether more than 1,000 planners who are fully involved in the whole process of research, design, review, construction guidance and post-implemention evaluation. They are an important force in solving “last-kilometer” problems.

附录 Appendix

中国历史年代简表
A Brief Chronology of Chinese History

远古时代 Prehistory			
夏 Xia Dynasty			c. 2070 - 1600 BC
商 Shang Dynasty			1600 - 1046 BC
周 Zhou Dynasty	西周 Western Zhou Dynasty		1046 - 771 BC
	东周 Eastern Zhou Dynasty		770 - 256 BC
	春秋时代 Spring and Autumn Period		770 - 476 BC
	战国时代 Warring States Period		475 - 221 BC
秦 Qin Dynasty			221 - 206 BC
汉 Han Dynasty	西汉 Western Han Dynasty		206 BC-AD 25
	东汉 Eastern Han Dynasty		25 - 220
三国 Three Kingdoms	魏 Kingdom of Wei		220 - 265
	蜀 Kingdom of Shu		221 - 263
	吴 Kingdom of Wu		222 - 280
晋 Jin Dynasty	西晋 Western Jin Dynasty		265 - 317
	东晋 Eastern Jin Dynasty		317 - 420
	十六国 Sixteen States		304 - 439
南北朝 Southern and Northern Dynasties	南朝 Southern Dynasties	宋 Song Dynasty	420 - 479
		齐 Qi Dynasty	479 - 502
		梁 Liang Dynasty	502 - 557
		陈 Chen Dynasty	557 - 589
	北朝 Northern Dynasties	北魏 Northern Wei Dynasty	386 - 534
		东魏 Eastern Wei Dynasty	534 - 550
		北齐 Northern Qi Dynasty	550 - 577
		西魏 Western Wei Dynasty	535 - 556
		北周 Northern Zhou Dynasty	557 - 581

<table>
<tr><td colspan="2">隋 Sui Dynasty</td><td>581 - 618</td></tr>
<tr><td colspan="2">唐 Tang Dynasty</td><td>618 - 907</td></tr>
<tr><td rowspan="6">五代十国
Five Dynasties and Ten States</td><td>后梁 Later Liang Dynasty</td><td>907 - 923</td></tr>
<tr><td>后唐 Later Tang Dynasty</td><td>923 - 936</td></tr>
<tr><td>后晋 Later Jin Dynasty</td><td>936 - 947</td></tr>
<tr><td>后汉 Later Han Dynasty</td><td>947 - 950</td></tr>
<tr><td>后周 Later Zhou Dynasty</td><td>951 - 960</td></tr>
<tr><td>十国 Ten States</td><td>902 - 979</td></tr>
<tr><td rowspan="2">宋 Song Dynasty</td><td>北宋 Northern Song Dynasty</td><td>960 - 1127</td></tr>
<tr><td>南宋 Southern Song Dynasty</td><td>1127 - 1279</td></tr>
<tr><td colspan="2">辽 Liao Dynasty</td><td>907 - 1125</td></tr>
<tr><td colspan="2">西夏 Western Xia Dynasty</td><td>1038 - 1227</td></tr>
<tr><td colspan="2">金 Jin Dynasty</td><td>1115 - 1234</td></tr>
<tr><td colspan="2">元 Yuan Dynasty</td><td>1206 - 1368</td></tr>
<tr><td colspan="2">明 Ming Dynasty</td><td>1368 - 1644</td></tr>
<tr><td colspan="2">清 Qing Dynasty</td><td>1616 - 1911</td></tr>
<tr><td colspan="2">中华民国 Republic of China</td><td>1912 - 1949</td></tr>
</table>

中华人民共和国1949年10月1日成立
People's Republic of China, founded on October 1, 1949

参考文献　Bibliography

法律、法规等（Laws and Regulations）

[1]　《保护世界文化和自然遗产公约》（"Convention Concerning the Protection of the World Cultural and Natural Heritage"）1972

[2]　《北京城市总体规划（2016 年—2035 年）》（*Beijing Master Plan* (2016–2035)）2017

[3]　《北京皇城保护规划》（"Conservation Plan for the Imperial City of Beijing"）2003

[4]　《北京历史文化名城保护条例》（"Regulations on the Conservation of the Historic City of Beijing"）2021

[5]　《北京市非物质文化遗产条例》（*Beijing Municipal Regulations on Intangible Cultural Heritage*）2019

[6]　《北京市古树名木保护规划（2021 年—2035 年）》（"Beijing Municipal Regulations on the Conservation and Management of Ancient and Valuable Trees"）2019

[7]　《北京市文物保护单位保护范围及建设控制地带管理规定》（"Beijing Municipal Regulations on the Management of the Conservation Scope and Construction-Control Zones of the Officially Protected Monuments and Sites"）2007

[8]　《北京中轴线文化遗产保护条例》（"Regulations on the Conservation of Beijing Central Axis Cultural Heritage"）2022

[9]　《公共服务领域英文译写规范》（"Guidelines for the Use of English in Public Services Areas"）2013

[10]　《故宫保护总体规划（2013 年—2025 年）》（"Master Plan for the Conservation of the Forbidden City (2013–2025)"）2017

[11]　《关于推动城乡建设绿色发展的意见》（"Guidelines on Promoting Green Development in Urban and Rural Construction"）2021

[12]　《关于在城乡建设中加强历史文化保护传承的意见》（"Guidelines on Preserving and Carrying Forward Historic and Cultural Heritage in Urban and Rural Construction"）2021

[13]　《历史文化名城保护对象认定标准与保护名录登录机制制定》（"Development of Criteria for the Identification of and the Registration Mechanism for Historic

Cities Under Conservation”）2021

[14] 《历史文化名城保护规划标准》（GB/T 50357-2018）（*Standard of Conservation Planning for Historic City* (GB/T 50357-2018)）2018

[15] 《历史文化名镇名村保护条例》（*Regulations on the Conservation of Historic Towns and Villages*）2017

[16] 《实施〈世界遗产公约〉操作指南》（“Operational Guidelines for the Implementation of the *World Heritage Convention*”）2021

[17] 《首都功能核心区街区保护更新导则》（“Guidelines for the Conservation and Renewal of Neighborhood in the Core Area of the Capital City of Beijing”）2020

[18] 《首都功能核心区控制性详细规划（街区层面）（2018 年—2035 年）》（“Regulatory Plan for the Core Area of the Capital (Block Level) (2018–2035)”）2020

[19] 《中国文物古迹保护准则》（“Principles for the Conservation of Heritage Sites in China”）2015

[20] 《中华人民共和国非物质文化遗产法》（*Law of the People's Republic of China on Intangible Cultural Heritage*）2011

[21] 《中华人民共和国文物保护法》（“Law of the People's Republic of China on Protection of Cultural Relics”）2017

著作（Works）

[22] 侯仁之 . 北京历史地图集（*Atlas of Beijing History*）. 北京：文津出版社，2013.

索引 Index

B

C

D